NO DUCKS IN THE ATTIC

& OTHER BASICS OF HVAC INSTALLATION

BY R. J. SCHUSTER

ISBN: I-4392-3204-0
ISBN-I3: 9781439232040

Visit www.booksurge.com to order additional copies.

FIRST THINGS FIRST

There are no "Ducks" in the attic, or any other part of the house. At least, there shouldn't be. What you may find are *"DUCTS"*. The first one has feathers and goes quack, quack. The second are rectangular and round tubes, usually made of metal or ductboard that transfer heated or cooled air from a source to a conditioned space (where the people are). There, now I feel better.

The purpose of this book is to establish a basic understanding of what should and shouldn't go on during a residential HVAC installation. After spending more than a decade in this industry, it became obvious to me that many technicians were not taught the basics of how, or why they do what they do. For many, the teachers they had may not have really known what was right, or just weren't able to explain it. For others, theory may have been taught in a classroom setting, but the practical experience that was delivered had all to do with how to braze, or cut, or fix something. This book will give you a different perspective than you have probably been exposed to before, and hopes to bridge the gap between duct wrangler and scwewy salesman. In any case, follow along in what I hope is an entertaining, yet informative look at how to successfully compete the task at hand. Whether you've been an installer for a day or a century, I'll bet you a pizza that you'll learn *something* that you never understood, or knew before. While this book is mostly geared to residential air conditioning, it can just as well be applied to boiler, furnace, or water heater installations. So, grab your beverage of choice, and let's begin.

CHAPTER I
Before You Start

Unlike the whole "chicken and egg" thing, before you even can start the installation of a new A/C system, we know that a sale has taken place, maybe by you, maybe by someone else, but somebody sold something. Central A/C systems these days aren't cheap. In fact, it's common to spend anywhere from $4,000 to $15,000 or way more to cool the average American home. Now it's your job to make sure that your customer gets what they paid for.

The details of what has been purchased by the customer should be spelled out as clearly as possible on a contract or work order. Included should be brand name, where the system will go, what areas are to be cooled, material specifics, who is responsible for cleaning areas such as basements or yards or any work area, repairing, patching, painting, framing, cutting, covering, upgrading, and virtually any activity associated with your project. Phrases such as "turn-key system" and "all inclusive" are often fine if those are truly your intentions, but it is far more important to clearly spell out what is not included. Those writing the contract always seem to forget that part. Something like framing in an exposed duct that wasn't included can cost the profit of the job, or adding four ducts to the lower floor that wasn't part of the agreement in order to get paid, is completely avoidable.

The contract should, of course, also include the price and terms of payment, along with any down payment, and financing applications. A capital improvement form for tax purposes (if necessary), and any permit applications may also be included. Most contracts have terms and conditions preprinted somewhere on the form which brings us to the most important detail; the signature. Without this you may complete the entire job to everyone's satisfaction, but when it comes time to get paid, you're a stranger, and you won't have a legal leg to stand on if they don't want to pay. So before anyone touches a tool, make sure ALL of the signatures are in place by the right people.

As the job is being assembled you should have a job folder of sorts that will not only contain the paperwork discussed above, but also a map with directions (if necessary), brochures if promised, rebate info, a material list of what you have and what you are missing, and a design layout and load calculations as necessary. Depending on the city and state you work in, you may need multiple permits or none at all. Play the game, pay the $25.00 or so, and make everyone happy. There's nothing more aggravating than having your local inspector watching everything you do, and telling you it's wrong after you did it, because you didn't feel like taking the twenty minutes to go to City Hall and fill out the form.

As your company (or the one you work for) grows you'll see that because you're becoming the bigger player in town, all eyes are on you. You may be encroaching on the local guys, so make sure you dot all the "I's" and cross all the "T's", so they at least can't complain about your work being illegal or shabby. Set the example. You may find that the workers at City Hall aren't even familiar with the permits you request, or have a different procedure each time you apply. Don't laugh; it's happened to me many times. If you spend some time with the person in charge, (the inspector or maybe the clerk) you may actually help them write the procedure and end up straightening out a mess they've been dealing with for years. That should change their attitude, if they ever had any doubt about your intentions. Help them; it's easier than trying to fight City Hall.

Back to the folder, it's also a good idea to have a list of possible add-on's, upgrades, or other sales that you may make during the course of the job. They may include anything from a better air filter or UV light, to an ERV (HRV), a light fixture in the attic, or even pull-down attic stairs. Ask your company for a list of prices and commission on these items.

If you're doing enough business, it won't be long before you need a skill that you don't have to complete the job so you'll need to make a choice. 1) Give that work away by telling the customer that they will have to handle that part, 2) become a jack of all trades, or 3) hire a subcontractor. Sometimes it really pays to reject that extra work and stick to what you're best at, depending on how difficult your customer may be. Having said that, assuming that you have, at least, reasonable customers and you have bills to pay, try the subs. If you get in over your head with a task you're not familiar with, chances are very good that you will end up spending all of the time and profit you would have made from the entire job. You probably know a whole group of trades' people who would appreciate the work. If you don't, there's always the Yellow Pages. Duct cleaners, electricians, plumbers, carpenters, painters, and others, all advertise there, and can even provide a great opportunity to increase your profits by marking-up their price.

You'll be amazed how a small duct job turns into a whole addition onto a house. I once had a client who was interested in putting A/C in his home, in which he also ran a small business. I knew he was doing O.K., he had about six people working for him in his garage turned office. Now this was a single raised ranch with a total of about 2,000 sq. ft. Normally, a basic A/C system in this house at the time would cost about $5,500. But it wasn't a basic system that he wanted. Not only did he go for the variable speed air handler, with two-speed condenser, he also insisted on four zones. You can't even imagine what the attic looked like. But that's not the point of this story. In one part of the house, we had to run two ducts exposed, which I agreed to box in (sub-contractor), then the paint didn't match, so we repainted the room at his request (sub-contractor), then the hall (painted to look like marble), and the entrance way (a custom sponge paint). Next, they wanted to upgrade the railing to the stairs, and the wife wanted to turn the spare bedroom into her closet, complete with built in shelves, bars, cabinets, island, and custom lighting. Of course then there wasn't enough room in the electrical panel, but we couldn't just add a sub-panel, they needed a service upgrade (sub-contractor). Naturally that required a service cable upgrade, but then we needed a new meter pan, which called for alterations in the vinyl siding, and finally they wanted all of the receptacles and switches changed in the entire house.

The final tally came to about $35,000, for which they happily wrote a check, and have called us back for more work, many times since. The best part is that even though more than half of the work was done by subcontractors working for me, there was a big, fat, net profit that far exceeded anything I could have dreamed of when I went to sell them a $5,500 attic A/C system. You can probably find work for subs in almost every house you are in, each time tagging on any-where from 10–50%. People are willing to pay a hefty premium for convenience, and if you, as an installer, can sell the additional work, you'll have a very nice income too.

Add-ons should be the first word on your mind between the hours of 8:00am & 4:00pm. There are so many ways for you to increase your income it's not funny. Discuss it with your manager; I'm sure they will be more than happy to pay you a nice commission on top of your pay. Your hourly rate will go up faster if your jobs are more profitable because you are selling the add-ons. Remember, you are already there anyway, and it probably won't add much time to the job to put in a better air filter, but I'll bet when you look at the profit on the job, it will have increased by close to 50% or more!

Remember: ADD-ON, ADD-ON, ADD-ON!

CHAPTER 2
Load Calculations

If you like to live on the edge of customer dissatisfaction, and you get all warm and fuzzy about the idea of doing things twice (or more), then just close your eyes, click your heels, and take a wild guess at the size of the units and ductwork for your next project. Or you can be like the few who actually try to do it right the first time. The way to do that begins with a load calculation. A load calculation is a process by which you determine the actual heating or cooling needs of the house. Years ago, they were done manually on paper using a "Manual J" book which gave you all of the charts and values you needed. Then, through numerous mathematical equations, you were able to determine the heat loss and heat gain of a structure. There are still a few hard core guys using the Manual J, but it is quickly becoming a lost art. Nowadays there are many, simple computer programs that make the task relatively easy. Most will print a report to refer back to if the homeowner or inspector has a concern about why you did what you did. In fact, many municipalities now require a load calc to be submitted with a permit application.

Load calculations will help you choose not only the size of the unit(s), but also the correct duct size for each room. While it's true that this is merely air conditioning for a house, and not rocket science, you still need somewhere to start. I know the other guy stands at the curb, holds up his right hand, closes one eye, and insists that because it's the second Tuesday of the month, his 1976 Nova started today, and that Joe's A/C Supply House is out of 3-ton condensers, this house is definitely a 3½ ton. Better yet, since he likes your dog, he'll "do you a solid" and make it a 4-ton. Just hide your daughters and run for the hills now, before he does you any other favors.

In all seriousness, there are a lot of reasons, besides the obvious, to do a load calculation. And, to be perfectly honest, some of you probably can come up with the right size units for the more common houses in your area. That being said, consider the following. Not only does doing a load calculation make you different from the competition, but also you will find that after you explain what it is that you are doing, your customer will feel like they are really getting their money's worth, and are less likely to question your decisions. I should also mention that there are now programs out there that allow you to do a basic load calculation on a PDA right in the home. This, in itself will be impressive to your customer. Still, these programs are limited in their abilities, and do not offer all the bells and whistles such as duct design, sizing, and layout. I believe that the traditional method will serve you better in the long run.

Start the process as you drive up to the house. Survey the outside; notice the layout, construction, how many stories, roof type and color. While you are at it, look at the location of the electric, the gas meter or fuel tanks, trees and shrubs, and any obstructions that may be a factor for the system layout.

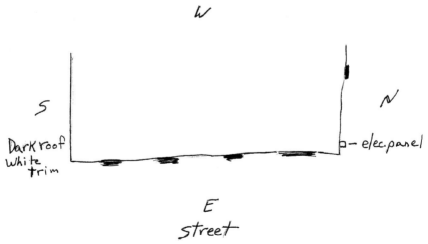

Fig I

Start a drawing of the house from outside. It's easier to get an idea of the footprint of the house, and any unusual offsets or additions, from outside. You'll also need to know what direction the house faces. Duct-work sizes can change significantly if a room with large windows and skylights faces south instead of north.

Once inside the home, as always, the first step is to put on booties or shoe covers. Explain to the homeowner exactly what your purpose is today. You are going to figure out, not only the size of the outdoor unit (the condenser), but also the amount of air flow (CFM) and BTUs needed for each room, so that you can properly size the air handler, and ductwork. Tell him or her that you'll need to measure each individu-al room and window. Other factors include hot tubs, computers, large screen TV's, number of people living there, and other heat producing items or activities that take place in this house. I always ask people how much cooking they do. Once a customer told me they only cook a few dinners a week, and that it was no big deal. Well during the course of the conversation over the next 20 minutes, the same customer who told me they didn't cook very much, also told me that they ran a catering busi-ness out of their home and do three or four hours of baking a day. Go figure.

After you do your brief explanation of what and why you're doing a load calculation, ask the homeowner if you can start here (usually the closest room), pointing to the living room. They say, "sure go ahead", and you're off. Now pull out the drawing you started outside, and add the two or three rooms you can see, to your drawing. Grab your tape measure, and walk to the edge of the room, and hook the end of the tape onto a radiator, doorsill, or crack in the floor. NEVER the furniture! If they see you do that, you just threw away all of your (and the salesman's) hard work about how you really do treat the house as if it were your own. This is very important. It may seem trivial, but it goes along with the whole first impression speech. While we are on the subject, don't put your clipboard, flashlight, or anything else on any furniture. Many people are very particular about their possessions. Even if your clipboard is made out of foam rubber and mink fur, and their couch is rusty cast iron, don't do it. Just don't. Treat the home with respect and you'll soon find the customer beginning to relax. This is definitely one area that you do want to over-compensate in. You'll often find then, that many people want to help you hold the end of the tape measure. I usually politely decline the assistance, explaining that I do this all day long solo. By yourself, the entire process in the house should take about 20 or 30 minutes. If you let them help, add another 20 minutes.

When measuring a room or window it's not necessary to measure in inches, feet are sufficient. A bedroom that is 10'3" x 14"9" should be referred to as 10' x 15', and a 42" x 58" window becomes 3½ x 5'. Remember, we're not sending this house into outer space, no need for a micrometer.

By now you're drawing should look something like this:

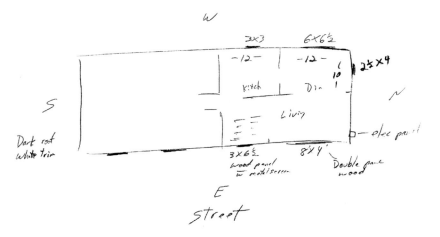

Fig 2

You'll notice that I always put window and door dimensions outside of the house, while the room dimensions go inside. Also, you should have noted direction of the house, where the electrical panel is, roof color, and trim color (in case you need to touch up the paint) including the leader pipe from outside the house. Continue through the rest of the house, asking permission to enter each room, and knocking on closed doors. When you come to the bathrooms, the next question that comes up is usually, "Oh do you put one in here too?" A good response is to explain that you definitely want to consider the bathroom when figuring out the size of the unit. As for whether or not you will put a duct in there should be determined by the load calculation. However, some people will tell you that they absolutely do, or absolutely do not want one in the bath. It's their house, and if they insist on something other than what you think is best, that's fine, but you must note it in the contract, and have them sign it. I think that a good policy is to not warranty any part of the job that you recommend against. Again, write it down and have them sign it. This is critical; don't downplay the documentation. I know a technician that says, "Write it as if you were writing it for a judge." Think about that.

Your drawing should be almost complete at this point. Be aware of what's on the ceiling such as lighting. Spot lights can give off an enormous amount of heat. Skylights are imperative. Carrying a ladder around isn't necessary, but you need to have a close approximation of the skylight size, and what direction it faces (east, south etc., not "up") and whether or not it's tinted. Notice also what's above the room. Is there conditioned space, or is it a roof, and if so, is there an attic or crawl space above the ceiling? Don't forget ceiling heights, which also add substantially to cooling capacity required. A room with ten foot ceilings has a full 25% more capacity than a room with eight foot ceilings; a detail that amateurs usually overlook.

While you are in each room, you need to note what heat producing equipment lives there, like computers, big screen televisions, extra lighting, and the like, and when you measure the windows make sure you look at the type of window (wood or metal, single or double pane, low E or tinted). Another item to look at is the floor. Is there carpeting there, or is it stone? Once again, just like the ceiling, is the space below the floor conditioned, or is it a concrete slab? Next, poke your head up into the attic for an R-value of the insulation, and while you are there, get a quick idea of how and where to set the unit, ductwork, condensate piping, etc.

Next, head down to the basement to look for insulation, and the depth below grade. While you're down there, check out the electric panel and anything else that may come into play during the project. If you are conditioning two floors with one system, you'll need to figure out how you can run the ductwork through the middle floor. In other words, if installing an attic system in a colonial, how do you get to the first floor? Or, if doing a basement system in a Cape Cod style house, how do you get the supply ducts up to the second floor? Don't forget the return duct(s), which are often even more challenging due to their size. For this reason, I usually put all of the closets on my drawing. That is, noting the closets on the drawing may help me to figure out where to run something when I'm back at the shop. It may be more difficult than you think, so plan ahead.

While we're on the subject, let me throw this common mistake at you. You can NOT heat and cool two floors with one system, on one zone. The second floor will always be too hot, and/or the first floor too cold. I know you've seen it before, or maybe even have done it yourself. If I could only put a number on the amount of time I've spent trying to get this point across, you would be stunned. Put your sales hat back on and go sell them a zone system. Don't let them tell you they don't want it. If they insist that they don't, document that they declined your recommendation on the contract or walk away from the job. Do the right job for your customer and save your reputation and your sanity.

Back to the ductwork placement; make sure you get the homeowner in on this discussion, because they absolutely need to know now what you're going to do, and what space you will need for the ductwork to properly complete the job. It's a mistake to start the project only to figure it out later. If you do wait, count on a heated debate beginning with "you didn't tell me you were going to do that! ".

Other items that are important to notice while in the house include how it looks when you first walk in. By that I mean, look at the curtains; are they open or closed? Are all of the neighbor's kids there running in and out of the house holding the door open for the parade of pets? What about the interior doors? Are the bathroom and bedroom doors open? Are all the doors closed with shag carpeting tightly sealing all the openings? However the house looks when you walk in is most likely how it always looks. These things may change your plans. Will you need a return in each room, or will one central return be enough? Take note.

Getting down on paper a sketch of what the house actually looks like isn't easy. The first house that you do might be a 30'x30' square, but for some reason your drawing looks like a map of South Boston. Don't worry; the drawing is only for you. No need to put your name on it and turn it in to the teacher.

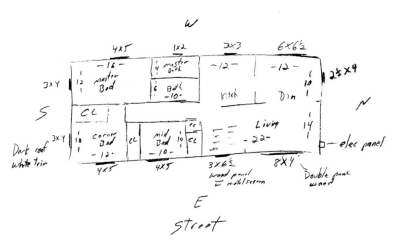

Fig 3

Okay, now what? You have your picture, and measurements, the trim color, and all your notes. How do we turn that into a cool house? Well, from here we move to the load calculation itself. Whether you do it manually, or by computer, you need to do it. As I said before, anyone who has been in the business for a few years, can probably take a good guess at the unit size, and be close enough for 60% of your customers. But who wants to go back and change the other 40% of improperly sized units or deal with unhappy customers? More importantly though, the load calculation is used to correctly size the ductwork, and determine how much air you need for each room. Everyone knows someone who "doesn't need all that stuff" and can do it by square footage. Hmmm, let's see now, room A is 15 x 20' (300 sq. ft.), and room B is also 15 x 20' (again 300 sq ft-DUH!). Room A belongs to a large 3500 sq ft house, and sits right in the middle of the structure, with no outside walls, and conditioned space above and below it. You may have guessed that room B is slightly different. In fact, room B is the end room of a one-story ranch. Three walls are outside walls, with east, south, and west exposures, minimal insulation, and three skylights with a total of about 60 sq ft. On the east and west walls, there are two windows each, measuring 4 x 5', and on the south wall, a 6 x 6 ½' ft. sliding door, and three more of these 4' x 5' windows. Just to top it off, they have thin curtains, if any at all, a large screen TV, two computers, and that is where the family of six chooses to hang out. Nope, I can't see any reason to use a highly technical, time-tested procedure, rather than "Voodoo Size-onomics" either!

There are many load calculation programs out there, with the Writesoft and Elite programs probably being the most popular. If you aren't currently using any software, talk to your local HVAC supply houses, check the trade magazines, or jump into the twenty first century and look on the internet (I promise it won't hurt), and you will surely find the latest product on the market. These programs will not only make a load calculation easy, but will also do energy comparisons, duct design, and a number of other interesting tasks. Some will even design the entire duct system, "blueprint" and all.

O.K., now you've got the latest and greatest. You are now ready to enter the information from your survey into the computer. These days, the questions are very direct, and anyone with minimal knowledge of residential air conditioning and basic construction will be able to breeze through it with a little practice. First you'll have to set the design standards, such as temperatures indoors and out. Outside design temperatures are published by the United States Air Force, and ASHRAE weather data, but you don't even have to search for them. They are likely already in your fancy new program; just find the pull down menu. These design temperatures will not be exceeded more than ½ % of the time. That means if the summer design temperature for your area is 88 degrees Fahrenheit, 99½ % of the time it will be below 88 degrees. This is important to understand so that you can explain it to your customer. What they specifically need to understand is that these numbers are not unlimited. If it's 110°F outside once every 18 years, we're not going to design the system to bring the house to 65°F. On that day, you *could* actually do that, but that would probably be the only day that the system works properly. The rest of the time it would short cycle, cool too fast, not dehumidify properly, be very uncomfortable and noisy, and shorten the life of the unit by years. Not to mention the ridiculous cost.

The accepted industry standard for the difference in the outside temperature as compared to the inside temperature in the Northeast is 18–22 degrees. In other words, when it's 95°F outside, for an extended period of time, you should be able to maintain 73-77°F inside. Keep in mind that it is never 95°F for an extended period; it almost always cools off at night. However, this requires your attention. Most people think that air conditioning systems have unlimited capabilities. They do NOT! This is the time to explain to the buyer that these systems are designed to make the house about 20 degrees cooler than the outside temperature. Now you can dazzle them with your knowledge of why you don't want to over or undersize their system like they may think (remember, short cycling, humidity, shorter compressor life, etc.?). Don't take any short cuts here. This is another discussion with your customer that is more important than you realize, and will save you many future phone calls.

Alright, back to the computer. Now you can enter all the room info including outside wall dimensions, insulation, number of people, extra BTU's for appliances, windows, construction type, and many other relative factors. Basically, just answer the questions. When you've completed all of that, you will end up with a few charts, maybe some pictures, and finally the meat of it all, the CFM and BTU numbers. Remember, you are sizing numerous items here, and not looking for, "Gee, it's a 3 ton." For the condenser, you should have something like this:

Sensible-	22,197	BTU's
Latent-	3,218	BTU's
Total	25,415	BTU's

As for the air handler or fan coil, on your load calculation you will find a CFM number like: 1456CFM (CFM being Cubic Feet of air per

Minute). We'll take a closer look at these numbers in the Equipment Selection chapter.

If you look a little farther, you should find a CFM number for each individual room. This is where the "built-in balancing" comes in, usually without the use of dampers. Hold on, don't get your socks all ruffled. I think dampers are very useful and can solve a whole host of problems, but I also know that they can be the cause of many unnecessary service calls. Whether the damper moved by itself, or someone just wanted to see if they worked, they will move at some point. So if you can achieve a balanced system by simply putting the right size duct in the first time, then some say you can save yourself a future headache. Remember, the more adjustments you give your customer to play with, the more times you're going to come back for free, and your customers will have the ill-conceived notion that your system doesn't work right, even if they are the ones who un-adjusted it! Do it right in the load calculation, tell the customer it's right because the computer says so, and they will brag about how high-tech your job was, and they won't dare disturb it. Anyway, that is why we want the CFM for each individual room, so we know what the airflow should be and we won't need dampers. However, if you do use dampers, make sure that you put a zip screw through the little hole in the handle to keep it in place, once you know it is properly adjusted.

At this point, take your sketch of the house, and in the center of each room, write the number of CFM required for that area. I like to put a circle around the number, so I don't confuse it with another mea-surement, because by now, you have numbers everywhere. This part of the exercise is really the most important part of the whole load calcula-tion, so pay attention, and put the numbers in the right places. Make sure you have CFM numbers for all the areas to be cooled, even if you don't plan to put a diffuser in that area. For example, maybe the mas-ter bedroom calls for 180 CFM, and the little 4 x 5 master bath only calls for 20 CFM. If you only look at the larger room, you may be between duct sizes, and not know which to use. However, when you include the smaller room as a part of the larger room, your decision will become clearer.

These silly little sketches can be very useful for, not only the load calculation, but also in the design work, and problem solving that you may do during and after the job. Additionally, I suggest that you make this a permanent part of the file for this customer for a few reasons. If you ever have a problem with the system, you'll know why you did what you did. Once in a while a homeowner will call you back six months later and say that one of the rooms is too warm. If you can go back with your sketch, you may find that when you did the work they had a little 2'x3' window on that south wall. But, alas, now there is a new 6'x9' sliding glass door on that wall that the customer didn't mention. If you forgot what that house looked like, you might have been convinced that you made a costly mistake. Luckily, you now have proof that you did the proper job at the time.

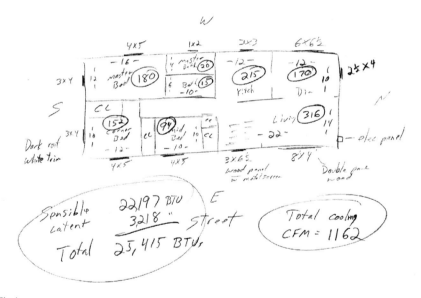

Fig 4

Now that you have all these numbers, let's turn them into some properly sized ducts!

CHAPTER 3
Duct Design And Sizing

Before choosing any equipment model or size, and before you start a material list, it's best to get an idea of the placement of the air handler and ductwork. Without this it's all just another guess. Where will the ductwork fit best, without adversely affecting airflow? Where will the return go? If you put the unit all the way over there, will it be easily serviced? You don't want to be the installer that everyone hates because you have to disconnect the return duct to remove the filter, or be a contortionist to access the service panel. Don't forget the drain lines and the refrigerant piping or line set. When it comes to piping, shorter is usually better. Hopefully you decided some of this stuff when you were doing the sketch of the house.

Now that you've decided on the placement of the air handler and ductwork, grab a blank piece of paper and draw a quick sketch of what it should look like when it's done. While it's true that some programs will do a duct layout for you, they do not take into account the real world obstacles like attic stairs, collar ties, etc.. You can print up a batch of templates to save a little time, if your company does a lot of similar houses.

Some of mine start out like this:

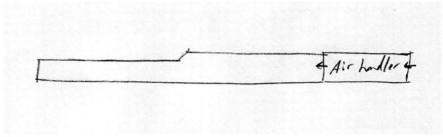

Fig 5

You may ask, "Why should I go through the trouble?" Well, there are many reasons to draw out the plan, as you'll soon see. For one thing, it's easier to make a material list when you can see it in front of you, and you won't have to figure it out three times because you forgot something. Or if one person does the load calculation and material list, and someone else is doing the installation, having a plan is very important. Without one you may make a material list for one design, and the installer may have a totally different idea for a layout, making your material list as useful as a fin-straightening tool on a fishing trip! You can also show how you planned to get around the non-movable obstacles you found. I like to put my layout in a plastic sleeve when it goes to the jobsite, so nobody gets ketchup on it.

Next, mark out the rooms on your drawing by referring back to your sketch. Look at the CFM required for each room, and determine what size duct you will feed it with. While there is some variation in airflow between hard duct and flex (flex delivering slightly less), in general the following applies when delivering air at a friction of 0.10 W.C., which is standard for residential applications (see your sliding duct calculator):

5" duct delivers about 50–60 CFM
6" duct delivers about 100–120 CFM
7" duct delivers about 150–165 CFM
8" duct delivers about 200–225 CFM
10" duct delivers about 380–420 CFM

Compare the numbers in the circles on your sketch to these to determine what size you will feed each room with. Also, you need to decide if a particular room will get one or two (or more) diffusers, such as the living room. The shape and length are often more of a factor when deciding how many diffusers a room will get, rather than CFM. If you have a room that is 15' x 24', and requires 430 CFM, sure, you could use one 10" run, but the air distribution, and comfort level would be greatly increased with two 8" runs. I have found that if a room is more than 16' long, performance and comfort are much better served with two or more diffusers to evenly distribute the airflow.

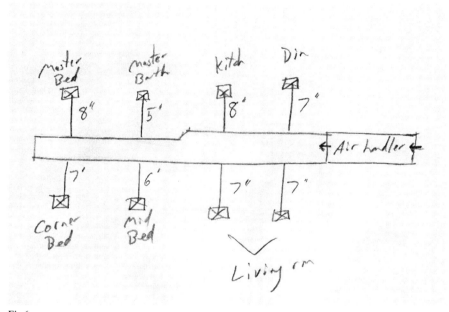

Fig 6

As far as residential duct systems go, you are much better off having a longer trunk line and shorter branches. Keep flex runs as even in length as possible, 12' or less as the space permits. A well-designed duct system will reduce its trunk size for about every four branch runs. This will insure steady pressure, and will help with keeping the system balanced. Imagine you design a rectangular trunk to run at 0.10" W.C. that will have 10 branch runs (all the same size for this example). After the first 5 ducts, you have roughly half the volume of air you started with, but the same size trunk. Without a reduction in the trunk size, the air from the branches at the end will have almost no velocity, or push, at all. Your customers will complain, "The other side of the house doesn't blow at all! "Your job is tough enough; let's not design problems into it. By the way, residential HVAC systems work best when supply air is designed to run at a pressure of roughly 0.10 inches of water column, and return air about half of that (0.05). Any more and the airflow will become noisy, any less and the house may be unbalanced.

Now that you have a system layout, and what size duct each room will get, you can figure out your trunk sizes, and how many lengths of each you will need. The supply side opening on some air handlers or coils are properly sized for your first section of trunk line, which will handle the total CFM of your system (1200 CFM in this example). Many are not, though, so don't assume because when you assume ... Decide which branch lines will come off of that first section of trunk line, and add them up. In this case it adds up to 720 CFM:

1–7"= 165 CFM Dining Room
2–7"= 330 CFM Living Room
1–8"= 225 CFM Kitchen
Total = 720 CFM

Take the total CFM of the system, here it is 1200, and subtract the CFM from the first section (720), and you are left with the CFM for the second section:

1200 Total CFM
−720 CFM for 1st 4 branches
480 CFM for remaining branches

Pull out the old duct calculator and find 480 CFM at 0.10" W.C., and match one of those sizes to your first size, and you now have a properly sized duct system. When choosing a size for the second part of the system, try to stick with a size that will match two sides of the first, larger duct size. In other words, if the first duct was a 12 x 18, don't choose an 8 x 14 for the second, because you would have to make two transitions. Choose a 12 x something, or an 18 x something and keep it simple!

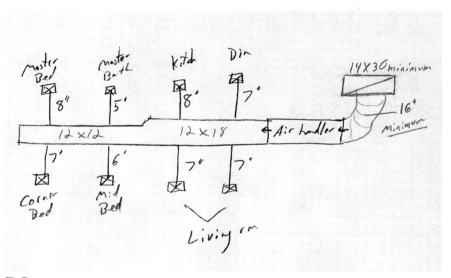

Fig 7

I'd like to go over one example that many installers seem to trip over. Design changes are more common on jobsites then they should be, and that's fine if they are done properly. Unfortunately, most of the time they aren't. Here is one problem I see all the time. The system below was meant to be a 3-ton (1200 CFM), *T-layout* such as this:

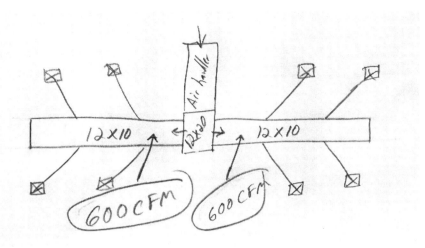

Fig 8

For whatever reason, it's decided that it would be easier to install as a straight trunk. So rather than asking for a new trunk, the unit is turned 90°, and the ductwork is joined end to end, or brought back to the shop and the system now looks like this:

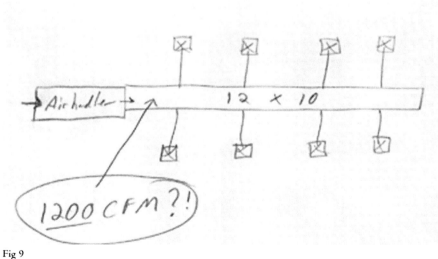

Fig 9

I hope you are noticing that the duct size was just cut in half. On the T, the 12 x 20 carried the total 1200 CFM at 0.1"W.C. It then went to **two** 12 x 10's, 600 CFM each. But then someone eliminated one of those 12 x 10's, effectively doubling the supply pressure. This is going to cause numerous problems including lots of noise, crazy airflow, an unbalanced system, and perhaps mountain goats to grow second heads.

The proper result with a straight trunk could have been achieved by adding one or more 12 x 20's before the 12 x 10 or even using what they had with a system looking like this:

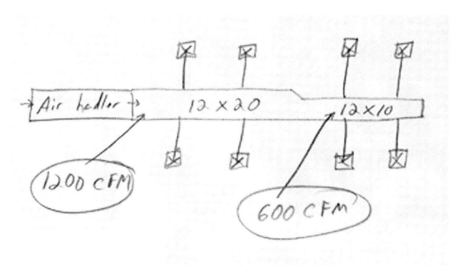

Fig 10

The most common problems that these changes cause are reduced airflow to sections of the house, and noise. All of which, of course, means hot and cold areas and hate mail to you.

Don't forget also, it is always better to increase the length of the trunk line than to increase the length of the flex or branch lines. I can tell you first hand that when those flex lines get to be much more than about 20 feet, you're going to have air flow problems. No doubt about it. Is it worth it? Save a half hour now, and have to come back on the first hot day? Forget it, it doesn't help anyone. Resist the pressure from the un-educated, because now you know better. While it's true that some guys over-engineer this stuff, some things *do* make a big difference. You just need to know which ones. Again it's those little details that make for a high quality, profitable job.

Since we have now spent some time on sizing the supply ductwork, let's go to the other end, the return. Contrary to popular belief, the return duct size is at least as important. Any air system can only deliver as much air as you give it. This is often overlooked when a problem arises. In fact, lack of return air is the sneaky cause of many problems common to the residential A/C system. I'm not the first person to go to a house where the owner complains that "the air blows really cold, but the house never gets cool enough", only to find a 5 ton system with two ten inch flexes for the return. (Go look at your "ductolator" again) Typical problems include return air noise, frozen coils, compressor failure, improper refrigerant pressures, insufficient cooling, premature blower failure, and very often condensate problems.

For now we need to be concerned with three aspects of the return system in our duct sizing and design, which are 1. Design pressure, 2. Size and, 3. Location. First, while the supply side is commonly figured to be 0.1" W.C., on the return side the goal should be somewhat less, *about* .05" W.C. Remember, you want to design a passive system, not one that will suck the napkins off the table. Second, once you know how much air you want to move 1200 CFM (3 Tons), for example, you can pull your trusty "ductolator" out again and match up 1200 CFM to the .05" W.C., and you will see that in round duct, you could use 17" duct, (which doesn't exist), and from experience, you will find that 16" will usually work fine, but 18" would be better. And third, we have location. Years ago it was an accepted practice to have a return in just about every room, and many companies still do it, and some houses still need it. However, it really isn't necessary in most situations for air conditioning. A central return in the hallway of an average ranch is sufficient. A central return in a custom, 4000 sq ft, 4 level house, with large windows, skylights, 25' ceilings, and air-tight rooms is surely a recipe for disaster. There is no exact formula for this, but you can start by watching for the following things. Whenever you have a cathedral ceiling, you definitely want to put a return there, and as high as possible, because that is where the heat is collecting. The same goes for sunrooms and additions. These rooms have much different heat gathering characteristics than the others. Capes (or Cape Cods) are notorious for collecting heat on the second floor. If you don't get return ducts to the second floor of a cape, I hope you like dealing with hot, irate customers. That's what you'll have when they find that, after spending all that money, it's still 15° hotter upstairs. If you want to stay in this business, find a way to get a return up there if it's the last thing you do before you meet your maker, or these jobs will **NEVER** end, and you won't get paid. Did I mention that these jobs would **NEVER** end? Anyway, you can never have too much return air, and you need to employ some common sense regarding placement.

These days, with energy conservation being the hot topic, we need to talk about duct insulation. Energy efficiency standards are constantly changing, but as we speak, supply ducts that run through non-conditioned space require R-8 insulation in most areas. Return ductwork may need anywhere from R-4 to R-8, so check your local code for verification. Aside from energy conservation, insulation is important on supply ducts on a/c systems for a number of reasons including sound suppression. More important yet, is the fact that when you run un-insulated metal ductwork through a damp basement the rain gods descend on the place and have a party.

Metal ductwork can be insulated internally, but that requires you to increase the size of the duct to compensate for it. Or, you can wrap the exterior with fiberglass insulation or reflective bubble-wrap. Ductboard comes in different thicknesses and R values. The one inch is much easier to work with, but will only give you about an R-4.5 which will then require you to wrap it anyway. Also, with metal duct, the inspector may require you to seal each joint with tape or mastic.

When your surveys, load calculations, design, and duct designs are complete, the next natural step is to start gathering material. Rather than just grabbing a bunch of stuff, let's be organized about it, and start with a list. Sounds simple, but after doing 10 or more, and always forgetting a good part of the job, you'll find that a pre-printed list is priceless. You may want to have separate lists broken up into categories such as A/C, electrical, furnaces, boilers, etc. A sample A/C list is shown below:

AC / FURNACE CHECKLIST   CUSTOMER _____

QTY	MATERIAL	REC	RET	PRICE
	FURNACE			
	CONDENSOR			
	AIR HANDLER			
	COIL			
	LINE SET			
	3/8 ACR COUP			
	3/8 ACR 90			
	ACR COUP			
	ACR 90			
	3/8 LIQUID DRIER			
	3/8 H/P LIQUID DRIER			
	CONDENSER PAD			
	HEAT PUMP LEGS			
	DRAIN PAN			
	4 X 4 'S			
	2 X 4 'S			
	PLYWOOD			
	JOIST HANGERS			
	BEAM HANGERS (WOOD)			
	UNISTRUT			
	3/8 THREADED ROD			
	3/8 NUTS AND WASHERS			
	3/8 ROD COUP			
	LEADER PIPE			
	LEADER STRAPS			
	1 1/2 GALV STRAPS			
	4" SMOKE PIPE			
	4" SMOKE ELL			
	5" SMOKE PIPE			
	5" SMOKE ELL			
	6" SMOKE PIPE			
	6" SMOKE ELLS			
	7" SMOKE PIPE			
	7" SMOKE ELLS			
	8" SMOKE PIPE			
	8" SMOKE ELLS			

QTY	MATERIAL	REC	RET	PRICE
	3/4 PVC PIPE			
	3/4 PVC TRAP			
	3/4 PVC TEE			
	3/4 PVC CAP			
	3/4 PVC COUP			
	3/4 PVC 45 DEGREES			
	3/4 PVC 90 DEGREES			
	3/4 PVC MALE ADAPTERS			
	X____ PVC WYE			
	CONDENSATE PUMP			
	CONDENSATE TUBING			
	2" PVC PIPE			
	2" PVC 45 DEGREE			
	2" PVC 90 DEGREE			
	3" PVC PIPE			
	3" PVC 45 DEGREE			
	3" PVC 90 DEGREE			
	DUCTBOARD			
	x____ DUCT			
	x____ DUCT			
	x____ DUCT			
	x____ DUCT			
	SHEET METAL			
	X____ METAL DUCT			
	X____ METAL DUCT			
	____ PLENUM STAND			
	FOIL TAPE			
	S-SLIP			
	DRIVE			
	ZIP SCREWS			
	18" COLLARS			
	16" COLLARS			
	14" COLLARS			
	12" COLLARS			
	10" COLLARS			
	8" COLLARS			

QTY	MATERIAL	RE	PRICE
	7" COLLARS		
	6" COLLARS		
	18" FLEX		
	16" FLEX		
	14" FLEX		
	12" FLEX		
	10" FLEX		
	8" FLEX		
	7" FLEX		
	6" FLEX		
	TIE STRAPS		
	14 X 30 STAMPED GRILL		
	20 X 30 STAMPED GRILL		
	X____ STAMPED GRILL		
	14 X 30 FILTER GRILL		
	14 X 30 FILTER		
	20 X 30 FILTER GRILL		
	20 X 30 FILTER		
	X____		
	X____ MEDIA FILTER		
	X____ FILTER RACK		
	MV-2		
	MV-3		
	MV-4		
	X__ X__ STR BOOTS		
	X__ X__ STR BOOTS		
	X__ X__ ANGLE BOOT		
	X__ X__ ANGLE BOOT		
	X__ X__ TOP BOX		
	X__ X__ TOP BOX		
	X__ X__ SIDE BOX		
	X__ X__ SIDE BOX		
	X__ X__ PENCIL BOOT		
	X____ WALL DIFFUSER		
	X____ WALL GRILLS		
	X____ FLOOR DIFFUSE		
	X____ FLOOR GRILLS		
	X____ CEILING DIFUSER		

Fig II

You'll notice that this form goes a little farther than just a basic material list. Every shop has a slightly different procedure, but this should be universally acceptable. Column 1 shows how many of an item will be needed. Column 2, how many were received by the install crew. Column 3 shows how many were returned from the job, and column 4, the total cost of all the items used on this job. Depending on what part of the country you work in, you'll need to adjust the list to your needs, of course.

This may seem like a trivial detail, the material list that is, but without these little details running properly, the big things will never be right. Without a procedure in place for these little details, a satisfying and profitable job is almost impossible. These are the details that will add 6 hours of labor to the job, which nobody will be able to account for. Can you tell that I think this is important? Try to assemble all the material, before the job starts, so that on the morning the job is to begin, the technicians have everything they need, and there is no confusion. Another possibility would be to deliver all the material to the home the day before the job starts. If your company is able to do this, give it a try. You may find that now your jobs are finished in 2 days, instead of 2 ½. And we all should know, there is no such animal as a "half-day"!

CHAPTER 4
Equipment Selection

If you look around at the different manufacturers, you will realize that there are probably as many variations in equipment as there are customers. Variable speed, two stage, 21 SEER, 90%, and these are only the tip of the iceberg when it comes to equipment selection. Some companies even offer different colors now! As for this discussion, we'll assume that the particulars of the system you are installing have already been determined by whoever made the sale. You only need to pick the right size, and don't forget the proper voltage.

You'll need your load calculation here again, and we'll look at BTU's first. The numbers you have should be broken down into sensible, latent, and total BTU's. Without getting too technical, sensible BTU's refers to the heat an object gives off. Latent BTU's refers to the amount of heat stored in an object, and total, obviously, is the combination of the two.

The idea here is to figure out what size condenser is needed to cool the house in question. Find the performance ratings for the model you will be installing. There you will see model number, SEER ratings, BTU's, and maybe a few other obscure things. Some manufacturers list sensible and total BTU's, others list only total. Whenever possible, use the sensible numbers from the manufacturer's literature, compared to the sensible numbers from the load calculation for a more accurate match. Most of us know a 3-ton unit to be 36,000 BTU's, or 12,000 BTU's per ton. However, if you actually look at the specs, you may be surprised in many cases, how the ratings vary from brand A to brand B.

Physical Data

Model	Condenser Fan Dia. In.	Condenser Fan RPM	Liquid Connection	Suction Connection	Type	Approx. Shipping Wt.
CK18-1	18	1060	3/8	3/4	Sweat	125
CK24-1	18	1080	3/8	3/4	Sweat	135
CK30-1	18	1080	3/8	3/4	Sweat	140
CK36-1/3	18	1080	3/8	3/4	Sweat	150
CK42-1	18	1080	3/8	7/8	Sweat	180
CK49-1/3	22	1080	3/8	7/8	Sweat	176
CK60-1/3/4	22	1080	3/8	7/8	Sweat	208
CK62-1	22	1080	3/8	7/8	Sweat	258

DIMENSIONAL DATA
CK18-62 MODELS

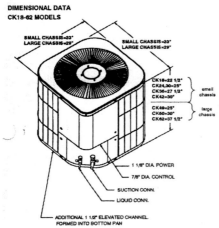

SMALL CHASSIS=23"
LARGE CHASSIS=29"

SMALL CHASSIS=23"
LARGE CHASSIS=29"

CK18=22 1/2"
CK24,30=25" } small chassis
CK36=27 1/2"
CK42=30"

CK49=25"
CK60=30" } large chassis
CK62=37 1/2"

1 1/8" DIA. POWER
7/8" DIA. CONTROL
SUCTION CONN.
LIQUID CONN.
ADDITIONAL 1 1/2" ELEVATED CHANNEL.
FORMED INTO BOTTOM PAN

Electrical Data

Model	Volts	PH	Wire Ampacity	*maximum Overcurrent Protection	Maximum Volts	Minimum Volts	Compressor RLA	Compressor LRA	Cond.Fan FLA	Cond.Fan HP
CK18-1	208/230	1	12.6	20	253	197	9.0	45	1.3	1/6
CK24-1	208/230	1	16.4	25	253	197	12.1	57	1.3	1/6
CK30-1	208/230	1	19.7	30	253	197	14.7	71	1.3	1/6
CK36-1	208/230	1	22.9	40	253	197	17.3	94	1.3	1/6
CK36-3	200/230	3	14.9	25	253	197	10.9	78	1.3	1/6
CK42-1	208/230	1	26.9	45	253	197	20.5	102	1.3	1/6
CK49-1	208/230	1	26.8	45	253	197	20.0	110	1.8	1/4
CK49-3	200/230	3	17.8	30	253	197	12.8	78	1.8	1/4
CK60-1	208/230	1	37.1	60	253	197	28.2	135	1.8	1/4
CK60-3	200/230	3	22.8	35	253	197	16.7	150	1.8	1/4
'60-4	460	3	12.1	20	506	414	8.3	73	1.8	1/4
62-1	208/230	1	40.8	70	253	197	30.8	147	2.3	1/3

*May use fuses or HACR type Circuit Breakers of the same size as noted.

Performance Ratings

Condenser Model	Evaporator Model	Total BTUH	Sensible BTUH	SEER	SRN/ BELS
CK18-1	AW18-XX	18400	12200	10.00	8.0 (4)
	AW24-XX	17000	12700	10.00	
	A18-XX	17000	12700	10.00	
	U18/UC18+EEP	17000	12700	10.00	
	A24-XX	17400	13200	10.00	
	U29/UC29+EEP	17400	12700	10.00	
	H24F+EEP	17400	13200	10.00	
	U31+EEP	18000	13100	10.50	
	A32-XX	18000	13500	10.50	
	AV18-30-XX	18000	13500	11.00	
	H36F/U32/UC32+EEP	18000	13500	10.50	
CK24-1	AW24-XX	22800	16800	10.00	
	AW30-XX/A24-XX	23000	17200	10.00	
	U29/UC29+EEP	23000	17200	10.00	
	H24F+EEP	23000	17200	10.00	
	U31+EEP	24000	18400	10.50	
	U36/UC36/H36F+EEP	24000	18400	10.50	
	H36F/U32/UC32+EEP	24000	18400	10.50	
	A32-XX	24000	18400	10.50	
	AV18-30-XX	24000	18400	11.00	
CK30-1	AW30-XX	27000	20000	10.00	8.0 (4)
	U29/UC29+EEP	27000	20000	10.00	
	A30-XX	27200	20500	10.00	
	U30+EEP	27200	20500	10.00	
	A38-XX	28000	21200	10.00	
	U31+EEP	28000	21200	10.00	
	U32/UC32/H36F+EEP	28600	21800	10.00	
	A32-XK	29000	21800	10.50	
	AV36-60-XX	29000	22100	11.00	
	U42/UC42/H49F+EEP	29000	22100	10.50	
CK36-1/3	U35/UC35+EEP	33000	24000	10.00	8.0 (5)
	U36/UC36/H36F+EEP	34000	25200	10.00	
	A36-XX	34000	25200	10.00	
	H49F+EEP	35000	26600	10.50	
	A42-XX	35000	26600	10.50	
	U42/UC42/H49F+EEP	35000	26600	10.50	
	AV36-60-XX	36000	27000	11.00	
CK42-1	U42/UC42+EEP	39600	27300	10.00	8.4
	A42-XX	40000	28000	10.00	
	H49F+EEP	40000	28000	10.00	
	U47/UC47+EEP	40000	28000	10.00	
	A48-XX	40000	29600	10.50	
	U49/UC49+EEP	40400	28200	10.00	
	A49-XX	41000	30400	10.50	
	U60/UC60/H60F+EEP	41000	30400	10.50	
	AV36-60-XX	42000	31200	11.00	
CK49-1/3	H49F+EEP	44000	33000	10.00	8.4
	U47/UC47+EEP	44000	33000	10.00	
	H60F+EEP	45000	34200	10.00	
	A48-XX	45000	34200	10.00	
	U49/UC49+EEP	45000	34200	10.00	
	U59/UC59/U60/UC60+EEP	46000	35000	10.50	
	A49-XX	46000	35000	10.50	
	H61F+EEP	47000	36000	10.50	
CK60-1/3/4	A60-XX	56000	40300	10.00	8.4
	U60/UC60/H60F+EEP	55000	39600	10.00	
	H61F/U61/UC61+EEP	56000	39500	10.50	
	U62/UC62+EEP	56000	39500	10.50	
	A61-XX	57000	42000	10.50	
	AV36-60-XX	57000	42000	10.70	
CK62-1	A60-XX	58000	45000	10.00	8.4
	U59/UC59+EEP	58000	42000	10.00	
	U60/UC60+EEP	58000	42000	10.00	
	U81/UC61+EEP	60000	43000	10.00	
	U62/UC62+EEP	60000	43000	10.00	
	H61F+EEP	60000	43000	10.00	
	A61-XX	62000	45000	10.00	
	AV36-60-XX	62000	45000	10.50	

1.) Note: XX Of A Model Designate Electric Heat Quantity.
2.) EEP - Order From Service Dept. Part No. B13707-38.
The Goodman Gas Furnace contains the EEP cooling time delay.
3.) When matching the outdoor unit to the indoor unit, use the piston supplied with the outdoor unit or that specified on the piston kit chart supplied with the indoor unit.
4.) SRN 7.6 with use of Sound Blanket accessory CSB-01.

Fig 12

ombination ratings continued

UNIT E-SERIES	INDOOR MODEL	TOT. CAP. BTUH	FACTORY-SUPPLIED ENHANCEMENT	STANDARD RATING	CARRIER GAS FURNACE OR ACCESSORY TDR†	ACCESSORY TXV‡	SOUND RATING (dBA)
	CK3BA036	33,400	TDR	10.50	—	10.50	82
	CK3BA042	33,400	TDR	10.80	—	10.80	82
	COILS + 58MVP120-20 VARIABLE-SPEED FURNACE						
	CJ5A/CK5A/CK5BA042	33,400	TDR	10.80	—	10.80	82
	CJ5A/CK5A/CK5BW036	33,400	TDR	10.50	—	10.50	82
	CK3BA036	33,400	TDR	10.50	—	10.50	82
	CK3BA042	33,400	TDR	10.80	—	10.80	82
	COILS + 58U(H,X)V060-12 VARIABLE-SPEED FURNACE						
	CC5A/CD5A/CD5BA036	33,400	TDR	10.50	—	10.50	82
	CC5A/CD5A/CD5BA042	33,400	TDR	11.00	—	11.00	82
	CC5A/CD5A/CD5BW042	33,400	TDR	11.00	—	11.00	82
	CD3(A,B)A036	33,400	TDR	10.50	—	10.50	82
	CD3(A,B)A042	33,400	TDR	11.00	—	11.00	82
	CD3CA036	32,400	TDR	10.50	—	10.50	82
	CD3CA042	32,600	TDR	10.50	—	10.50	82
	CD5A/CD5BW036	33,400	TDR	10.50	—	10.50	82
	CE3AA036	33,400	TDR	10.50	—	10.50	82
	CE3AA042	33,400	TDR	11.00	—	11.00	82
	CJ5A/CK5A/CK5BA036	33,400	TDR	10.50	—	10.50	82
	CJ5A/CK5A/CK5BN036	31,400	TDR	10.50	—	10.50	82
	CK3BA036	33,400	TDR	10.50	—	10.50	82
	CK3BA042	33,400	TDR	10.80	—	10.80	82
	COILS + 58U(H,X)V080-16 VARIABLE-SPEED FURNACE						
6-30, 31, 32, 33, 50, 52, 62	CC5A/CD5A/CD5BA036	33,400	TDR	10.50	—	10.50	82
	CC5A/CD5A/CD5BA042	33,400	TDR	11.00	—	11.00	82
	CC5A/CD5A/CD5BW042	33,400	TDR	11.00	—	11.00	82
	CD3(A,B)A036	33,400	TDR	10.50	—	10.50	82
	CD3(A,B)A042	33,400	TDR	11.00	—	11.00	82
	CD3CA036	32,400	TDR	10.50	—	10.50	82
	CD3CA042	32,600	TDR	10.50	—	10.50	82
	CD5A/CD5BW036	33,400	TDR	10.50	—	10.50	82
	CE3AA036	33,400	TDR	10.50	—	10.50	82
	CE3AA042	33,400	TDR	11.00	—	11.00	82
	CJ5A/CK5A/CK5BA042	33,400	TDR	10.80	—	10.80	82
	CJ5A/CK5A/CK5BW036	33,400	TDR	10.50	—	10.50	82
	CK3BA036	33,400	TDR	10.50	—	10.50	82
	CK3BA042	33,400	TDR	10.80	—	10.80	82
	COILS + 58U(H,X)V100-20 VARIABLE-SPEED FURNACE						
	CC5A/CD5A/CD5BA036	33,400	TDR	10.50	—	10.50	82
	CC5A/CD5A/CD5BA042	33,400	TDR	11.00	—	11.00	82
	CC5A/CD5A/CD5BW042	33,400	TDR	11.00	—	11.00	82
	CD3(A,B)A036	33,400	TDR	10.50	—	10.50	82
	CD3(A,B)A042	33,400	TDR	11.00	—	11.00	82
	CD3CA036	32,400	TDR	10.50	—	10.50	82
	CD3CA042	32,600	TDR	10.50	—	10.50	82
	CD5A/CD5BW036	33,400	TDR	10.50	—	10.50	82
	CE3AA036	33,400	TDR	10.50	—	10.50	82
	CE3AA042	33,400	TDR	11.00	—	11.00	82
	CJ5A/CK5A/CK5BA042	33,400	TDR	10.80	—	10.80	82
	CJ5A/CK5A/CK5BW036	33,400	TDR	10.50	—	10.50	82
	CK3BA036	33,400	TDR	10.50	—	10.50	82
	CK3BA042	33,400	TDR	10.80	—	10.80	82
	COILS + 58U(H,X)V120-20 VARIABLE-SPEED FURNACE						
	CC5A/CD5A/CD5BA036	33,400	TDR	10.50	—	10.50	82
	CC5A/CD5A/CD5BA042	33,400	TDR	11.00	—	11.00	82
	CC5A/CD5A/CD5BW042	33,400	TDR	11.00	—	11.00	82
	CD3(A,B)A036	33,400	TDR	10.50	—	10.50	82
	CD3(A,B)A042	33,400	TDR	11.00	—	11.00	82
	CD3CA036	32,400	TDR	10.50	—	10.50	82
	CD3CA042	32,600	TDR	10.50	—	10.50	82
	CD5A/CD5BW036	33,400	TDR	10.50	—	10.50	82
	CE3AA036	33,400	TDR	10.50	—	10.50	82
	CE3AA042	33,400	TDR	11.00	—	11.00	82
	CJ5A/CK5A/CK5BA042	33,400	TDR	10.80	—	10.80	82
	CJ5A/CK5A/CK5BW036	33,400	TDR	10.50	—	10.50	82
	CK3BA036	33,400	TDR	10.50	—	10.50	82
	CK3BA042	33,400	TDR	10.80	—	10.80	82
042-30, 31, 50, 51, 61	CC5A/CD5A/CD5BA042*	40,000	NONE	10.00	10.10	10.10	82
	CC5A/CD5A/CD5BC048	39,500	NONE	10.00	10.10	10.10	82
	CC5A/CD5A/CD5BW042	40,000	NONE	10.00	10.10	10.10	82
	CC5A/CD5A/CD5BW048	40,000	NONE	10.00	10.10	10.10	82
	CD3(A,B)A042	40,000	NONE	10.00	10.10	10.10	82
	CD3(A,B)A048	40,000	NONE	10.00	10.10	10.10	82
	CD3CA042	38,000	NONE	—	10.00	10.00	82
	CD5A/CD5BA048	40,000	NONE	10.00	10.10	10.10	82
	CE3AA042	39,500	NONE	10.00	10.10	10.10	82
	CE3AA048	40,500	NONE	10.00	10.10	10.10	82
	CG5AA048	40,500	TXV	10.00	10.10	—	82
	CJ5A/CK5A/CK5BA042	40,000	NONE	10.00	10.10	10.10	82
	CJ5A/CK5A/CK5BA048	40,000	NONE	10.00	10.10	10.10	82
	CJ5A/CK5A/CK5BN042	39,000	NONE	10.00	10.10	10.10	82
	CJ5A/CK5A/CK5BN048	39,000	NONE	10.00	10.10	10.10	82

e notes on pg. 18.

Fig 13

Over sizing a condenser can cause many performance problems, including not removing moisture in the house, and shortened condenser life due to short cycling. Short cycling, if you're not sure, is when an oversized unit heats or cools an area too quickly, therefore turning on and off more frequently than it is designed to. If a manufacturer engineers a piece of equipment to run an average of fourteen minutes per cycle in a properly sized house, but yours only runs three minutes per cycle, the system is oversized, short cycling, and headed for an early demise. So you want to be careful not to oversize, but under sizing is not correct either. The result will be a house that won't cool quickly enough (if at all), and equipment that will waste energy and never shut off. The practical solution here is to stay as close as you can to the ratings, which are never perfect, and if you can't decide between two sizes, err to the larger.

When you are ready to choose an air handler, don't assume that because the condenser was a particular size, the indoor unit will be the same. Very often it is not. Go back to the load calculation again, and find the total CFM needed. Then find the air handler section of the manufacturer's book, and look for the match. This may be tricky. The book will show you, not only model numbers, but also voltage ratings (208/230) and the performance in each fan speed (usually low, medium, and high) for each voltage. Again, we try to design residential systems at 0.1" W.C., so check the chart, and you should find different pressures also. With air flow, 400 CFM is considered 1 ton, and when you start looking at all of these charts, you'll realize that a 3 ton air handler doesn't necessarily mean that you get 1200 CFM out of it, especially when you get to the bigger sizes. Another quick note about evaporator coils (yes, every air handler has one). You can use a coil/air handler larger than the condenser, such as a 3 ½ ton coil, with a 3 ton condenser, in fact in some cases it's required in order to achieve the desired SEER rating. However, you can NOT reverse that idea. By that I mean, a coil sized smaller than the condenser will likely freeze and resemble the polar ice caps, reducing your air flow to zilch.

As you try to make your decision on a properly sized system, you will see that you could use, let's say for instance, a 2 ½ ton air handler on high, a 3 ton on medium, or perhaps a 3 ½ ton on low. Which should you choose? Well, being in the middle of the road, in many cases, may mean that you can't make a decision. Not this time. The middle is where you want to be, medium speed. Why? Well, a few reasons. First, if for some reason the customer complains about noise, or wants the system to run longer, so they can filter the air more efficiently, you can drop the speed to low. If they ever decide to finish the room in the basement, for Uncle Charlie with the peg leg, your system will have a little extra capacity by switching the fan to high. *(Keep reading for a note of caution on high fan speed).

As we all know, history seems to repeat itself. Undoubtedly, at some point you'll have a job that was sold for a little less than it should have. Or, maybe someone missed something during the planning stage that makes it a so-so job, profit-wise. You then may start to look for ways to cut corners. Although I don't recommend cutting corners to begin with because it always comes back to bite you, if you must, don't do it with air handler sizing. By that I mean installing the smaller unit, and running it on high. Pay attention now, and I'll tell you why. First of all, the amount of money you'll save with the smaller unit is negligible. More importantly, did you know that many air handlers, when run on high speed, will develop a negative pressure and create condensate problems, even without a filter? What I'm talking about here, is that on the high fan setting of some air handlers the water in the condensate pan will actually not drain because the fan is sucking air up through the drain piping! I'm sure a lot of you think that I'm pulling your femur (leg bone), but it's a fact. Why would a manufacturer sell a product that cannot run on one of the settings they give you without having a major problem? I dunno, but they do it! I can direct you to some trashed ceilings to prove it. Finally, after some time, I noticed that a few of the manufacturers actually warn about this condition in the installation manual. I could have saved myself a lot of trouble if I didn't already "know it all"!

Few installers read the installation manuals provided with the equipment they wrestle with. I know, most of you probably aren't big readers to begin with. I know I'm not, in fact I dreaded having to go back and edit this book because I don't like to read much. But, since they are paying you to do this job, you might as well try and get it right. You certainly don't have to read every word in the manual, and there are lots of things in there that really will make your job easier. If I were you, I would tell my helper to bring all the stuff in the house and get set up while I study up on the details. See, already the book is making your job easier! Do a little reading and you'll be surprised at what you can learn.

CHAPTER 5
Jobsite Set-Up

"You never get a second chance to make a…blah, blah, blah. Who cares? They're not even watching us! Later for that 'it's show time' stuff, it's too hot out for that".

From the second you start your truck someone *is* watching you. It may be your customer, it maybe Mrs. Jones across the street, or maybe just the crazy WWII pilot that waves to everyone on the corner, but they are watching. How you first appear to your audience will set the tone for the next few days on this job. All the little details of how you look and act are very important, starting with your arrival and set-up.

First, as you turn onto your customer's street, try not to run over any bicycles, or knock down any mailboxes. Find a place to park that will be convenient to the jobsite. No, not the lawn. Hopefully, there is a driveway you can park in. Ask the homeowner for permission to do so before you start, and watch out for the basketball hoop (experience talking). As you meet the homeowner, pay close attention to your personal appearance. It's certainly understandable if you are a little disheveled looking after spending 8 hours in an attic, but at 8:00 AM, you should look like you just left the dry cleaner. If you care about your appearance, the customer will feel much better about having you in their house. This means a complete uniform, clean-shaven, neat hair, shirt buttoned and tucked in, belt, black work boots (never sneakers, or ducky boat shoes), nametag, etc. In fact, you should all look like clones. After all, this is a business, not a dance club.

As you enter the house, put on shoe covers or "booties", even if your host says it's not necessary. The same goes for the drop cloths.

Even if the floors in your truck are nicer than their carpet, and many times that's the case, you will build tremendous value with that simple act. Now comes the hard part for some reason. Get *more* drop cloths. That's right, cover the *entire* floor from where you enter the house, to where you will be working, and any other entrance you may use. Of course, I'm not talking about the garage or other unfinished area, but all the finished floors. I once heard of a customer who asked when the crew would be completed with their work so that she could schedule the carpet cleaners, because she knew that they would make a mess. That kind of crap drives me crazy. Anyway, don't just put down a drop cloth so you can tell your manager that you did, do it so the customer realizes that you respect their home, and also to reduce complaints. You'll be amazed at how many fewer problem customers you will have, just because you went through the process of covering the floors. Suddenly they'll overlook the little mistakes because you tried to be careful, instead of blaming you for things that you were never near.

Once in the house, survey the area for any potential hazards, like furniture, paintings, babies, or mirrors that may be in your path. Move them now before there is even a question of damage (have the homeowner move the babies). What about outside? Are there toys strategically placed for your careful demise? How about one of those basketball sized, colored, glass balls on a pedestal in the middle of the lawn? If you're interested in one, I think you can find them in the "101 Ways to Get a Discount from Your Contractor" catalog.

O.K., back in the house, take a walk around the premises with the homeowner, and go over everything you can think of, from where the air handler and ductwork will sit, and where the grills and diffusers will go, right down to height of the thermostat. Don't forget the exposed duct in the closet, and the bush that will need to be trimmed for the condenser. Pull out the contract, and make sure everyone is on the same page. Whose responsibility is it to frame in that duct, and sheetrock, and paint it? It may seem obvious to you, but if it isn't in writing, you can be 99% sure, that before you get paid, you'll be the new painter in town.

I recommend taking the customer out in the driveway, and showing them the equipment that they are spending so much of their hard earned money on. Reinforce the sale right there; tell them what a good choice they've made. Show them the label, model number, SEER rating, and anything else you can think of. Every item you go over with them now is one less possible complaint later. Just like with the drop cloths, you're bringing the customer onto your side of the fence. You're now on the same team. SCORE!

At this point, ask your host for a staging area. This will be your temporary shop away from your shop. Hopefully, they will allow you to use a garage or similar area. Unload everything from your truck that you will need for this job, then organize, and stack it neatly in the same basic location on every single job. Flex on the far right, electrical stuff next, etc. Soon you'll see that with a lightning quick glance, you, and your superhero status, will be able to pick out any screw, or strap that may be absent from your inventory. Compare what you have, to the list of material you should have gotten from a supply house, or your shop. Start a list of additional things you will need to complete this project properly, and call it in <u>once</u>.

Unless your assistant is an actual Billy goat, you will need a place to store your trash. (I'm going to make believe I didn't hear that comment about leaving the trash at the curb!) The empty box from the air handler is a favorite. Use it! This is really another area where you can make or break people's impressions about you and your company. Look around, would you want YOU working in your house?

We all know how hot it can get in an attic in July, so plan ahead, and wear the appropriate clothing. Stay in uniform at all times. Check with your supervisor for hot weather uniforms. There are plenty of ways to help stay cool, such as working earlier hours, portable a/c units, and even ice vests. Of course, bring and drink plenty of fluids. Heat stroke is not something that just happens to the other guy. Working in a 140 degree attic for any length of time is serious business, so take serious precautions. Talk to the guys in the supply houses, they probably have something that can help.

Now for a few no-no's. Don't lose the shirts, ever. Don't use their pool, or their tools or anything unless you absolutely have no choice, and even then, don't do it. It's unprofessional, and will be perceived that way. Don't wander around areas that you're not working in; it will make them nervous or suspicious. Smoking is another touchy area. If you must smoke, do it in your truck, and leave the butts in the truck. Under no circumstances is it ever acceptable to smoke in the house. I don't care how many cartons a day they inhale. Remember, for the homeowner, this is not a comfortable time. Having strangers working in your house is a very personal and invasive thing. Many people become defensive just because you're there, even if you didn't do anything yet, so keep that in mind and respect their property.

You're just about ready to start now, so set your plan. The most efficient jobs are just that because everyone knows what their job is, and what comes next. The individual job itself may dictate the order of events for any number of reasons, such as, it's too hot in the attic after 1:00, or the baby takes a nap at lunchtime. So, plan accordingly.

If you are working on an attic system on a hot summer day, you may want to change your regular plan, and set the air handler first, and then run the electric, so that you can get some air moving up there as soon as possible. There really isn't any right or wrong way to do it, other than the fact that you need to have a plan, and everyone working on this job should know their role. Try to set a schedule or time limit for each task. Make a game out of it. Just by being aware of how long something took last time, you will automatically find yourself trying to beat your last time to keep it interesting. There are no rules that require installations to be boring, except, maybe in Kansas.

CHAPTER 6
Tools

Any person you know, that has a job, uses some kind of tool. Everyone works with tools. Chefs use pots and pans, spoons and spices. Doctors use stethoscopes and things to stick you with, and moms use diapers. But the tools of the HVAC professional installer (that's you) are more like, well, tools.

Auto mechanics buy tools like Imelda Marcos buys shoes. I don't know why. Maybe it's the chrome or something. If you ever talk to one, while they are near their collection, ask to see it (the toolbox, that is). They'll get all excited, unlock it, and even set up a ladder so that you can see inside of it. Some of these toolboxes even come with built in stereos! What you will undoubtedly find is that not only haven't they used 85% of all those shiny things, (imprinted with Jeff Gordon's signature), but they don't even have any idea what they're used for! Well, save your money boys and girls, because you don't have to live in your toolbox like they seem to. Most of the tools that you buy, and that you will use on a daily basis, can fit in a bucket or a small tool bag. One of the nice things about being an installer is that you really don't need to spend a ton of money to get going. Sure, there are a few larger items, like a cordless drill, Sawzall, etc., but your shop may provide many of these items, or maybe they have a tool account for you. The point is, you really only need the basics to start. Keep in mind, there will be many items dropped into wall cavities, or left behind in attics, so if you need to get emotionally attached to something, get a Cabbage Patch doll, or go into auto repair!

As time goes on, you will discover, however, that there are some things that will make your job easier and less frustrating. I'm all for spending money on tools such as an electrician's fish tape, or a mini brake, and a sheet metal hole cutter. And one of the best investments you could make is a cordless circular saw; just don't take it to bed with you. Remember, after all, that this is your career, so make sure you have what you need to do the job properly, quickly, and as easily as possible.

Oh, one other thing, it doesn't matter how good your eyesight is, or how straight you can hold a tape measure, you MUST own, and USE both a level and a square! Act like a professional, remember, perception is reality!

Here is a quick list of basic install tools

Drop light	**Cordless drill**
Sawzall	Dykes
Extension cord	PVC saw
Wire strippers	Level
Tape measure	Gloves
Kneepads	Hole saw kit
Channel lock pliers	Keyhole saw Level
Square	6 in I screwdriver
Drill bits	Pilot bit (18")
Tubing cutter	**Tubing bender**
Squeegee	Duct board knife
Schrader pin remover	Safety glasses
Torch	Electric tester (simple)
Hammer (claw, and lump)	Sheet metal sheers (straight, left, right)
Fire Extinguisher	

The list doesn't end here, though. There is always another crimper or cutter you can add to your collection. Remember though, you don't need to spend all of your hard earned cash right away. You'll probably be in this business for a while, so take your time, and wait to see how everyone else tackles these projects. You may find that your mentor uses something that you don't like, and there is a better tool for the job. Don't skimp on the safety stuff, though. Get a good pair of protective glasses, etc. You are responsible for your safety first.

CHAPTER 7
Not My Job, Man

Who's driving this train, anyway? Well, there is definitely an owner, but after that there are many variations depending on the particular shop. You may have some, or all, of the following positions in your company; general manager, controller, department head (installation manager), field supervisor, salesman, technician, apprentice, helper, etc. The same person may actually fill many of these positions. Whatever the hierarchy of your company, it is essential that everyone know what is expected of them, and where their responsibilities lay. Whatever your position is, you need to be able to do it well, without any thought. Get so good that you can do it in your sleep. I don't care if it's folding the drop cloths, or brazing in a tight space. If your company is going to count on you, you must be able to produce consistency.

Consistency is a key element to profit in any business. No profit; no work! You can't get more basic than that. The most valuable employees are usually the ones who are the most predictable so they can be assigned the proper jobs, and execute them without six phone calls, 30 questions, and 4 hours of overtime. Your company should be a well-oiled machine, and you are an important part of it. The only way to get that good at something, after the proper training, is to do it over and over and over again. Helpers and/or apprentices are often anxious to move up and learn more, which is great. However, they all too often get what they ask for, and end up failing at their expanded responsibilities because they really weren't ready yet. Anything worthwhile takes time, and there is no substitute for practice and experience, so don't rush it.

Look at the military. Do you think that they run all those drills and training missions because they have nothing else to do? Of course not! They do them so that when the chips are down, and their lives are on the line, nobody forgets what to do. They fall right into autopilot and don't even have to think about what needs to be done. Luckily you shouldn't have to endure any rocket attacks, (although some customers might make you think that would be easier than putting up with their B.S.)

The same ideas work in the real world, too. A technician and apprentice should be paired up for extended periods of time, so that you know what to expect from each other. Of course, it works best if your company offers an official training program for apprentices, so they would all be interchangeable. That way, whoever was matched together on any given day, could jump into any given truck, go to any job, and everyone would know exactly what to do. If only!

You will have to put on the show every day. The more organized you appear to both the customer, and your supervisors, the easier your life will be. Take this to heart, because this is real. Even if you don't *feel* organized, if you look and act like you are, you will reap the benefits. <u>Perception is reality!</u> Remember, people believe what they see.

One other word I want to discuss here, before I get into specific responsibilities, is pride. Pride in our own work doesn't seem to be as prevalent as it was in years gone by. Now everything seems to be based more on quantity, rather than quality, but just because you have a deadline to meet, doesn't mean you can't do it with pride. Be proud of the job you do, not only when you are done, but while you are doing it, too. Your work is an absolute, direct reflection on you. When the neighbors on the jobsite walk by, what are they thinking? Are you acting in a professional manner? I can tell you for sure, if you received this book or course from your employer, they care a great deal about the image you project. Look around you, would you want you or your partner working in *your* home?

Okay, enough preaching for this chapter. Now we're going to talk about specific responsibilities for both the tech, and the apprentice or helper. As previously mentioned, cleanliness and organization are enormously important for the technician or field person, but are even that much more so for the installation crew. Why you ask? Well, for one thing you don't carry your job into the house in a box. Yours comes in 20 boxes! And you won't be gone by 11:00 AM. You may be there 11 days. The usual response around this time is, "You're right! I/we need to be more organized!" Great! Admitting your problem is the first step to recovery. But, while many installers know they need to improve in this area, it usually doesn't change. I'll tell you why. Right now, what's probably running through your mind is, "He's right! I gotta get that dumb helper/tech to change everything he's doing wrong." The big revelation here is that each <u>team</u> (new concept) needs to establish who will be responsible for which tasks, and in what order. As simple as this sounds, it is rarely done. Everyday frustrations run rampant for helpers, because they are inexperienced and probably new to the industry, or at least to your company's way of doing things. They don't know what's expected of them, because nobody ever told them. Well, now I'm telling you.

***

The Apprentice/Helper

The person in this position generally does not have a whole lot of experience, if any, in this field. Regardless of that fact, the apprentice has many important jobs. I don't say this to make you feel all warm and fuzzy and needed, your wife or girlfriend can do that for you. You, as the apprentice can either make the job safe, easy, and organized, or you can make it a big pile of cuss words. Let's start with the truck. While this is ultimately the responsibility of the tech, he's got enough to worry about. The truck is like your shop away from the shop. On it, you've got tools, equipment, parts, drop cloths, and other things necessary for most jobs. The day should start with you, the apprentice, going through the truck, cleaning it out, organizing tools and parts, making sure you have a good supply of everything from booties and drop cloths, to a full B-tank, and zip screws. You should never have to waste time on a jobsite cleaning a tool, or searching for a part. This is real basic stuff, but it's these basics on which your whole career builds. Next make sure you are loaded up with the day's material (check the material list…Do it anyway!), and that the truck itself has enough oil, water, etc. Now on to the jobsite!

Once you are on the jobsite, have met the owner, and established a staging area, pull out the drop cloths, runners and booties first. While the tech is checking out the house and going over things with the homeowner, you should be covering the entire path from the front door, to wherever it is that you will be working. Move furniture, or anything else that could possibly be damaged (make sure to ask the homeowner first, before you actually move anything). Don't skimp with the drop cloths, the more the homeowner thinks you care about their home, the better your overall experience will be while you are there. Remember, <u>perception is reality</u>. You'll see!

When you are done with the big cover-up, don't waste time following the tech and the homeowner around the house, that's not your job. Start unloading all the material and tools you will need here, into the staging area. This is the next area you can probably improve quite a bit. Try to organize everything in some type of order, and do it the same way on every job. All the flex neatly stacked on one side, all electrical stuff laid out neatly on the floor, out of the way, so that everyone can see it, but won't trip over it. Do it the same way on each job so when someone is looking for the zip ties they know they should be right next to the flex, or wherever you always put them. Now start a garbage box or bag. I know this sounds basic, but believe me, it needs to be spelled out, and enforced. It's a good idea to have these responsibilities written out and mounted somewhere, so that you don't have to ever waste time waiting for directions.

***

After the set-up is complete, the technician should establish a procedure for this job and let you know if he wants you to work on something by yourself, or work with him. If it's with him, you should always try to anticipate what he is going to ask for next, whether it is a tool, material, B-tank, or whatever, try to stay a few steps ahead.

As far as discussing the job with the homeowner, it is important that you don't. The tech will tell them whatever it is that they need to know. I'm not saying that you shouldn't speak to them at all, but more often than not, helpers try to appear more knowledgeable about something than they really are. You may end up either contradicting the salesman, or tech, or say something that is altogether wrong. Leave all the business talk to the tech and salesman, and you'll lead a much happier life, believe me. Plenty of helpers have said something contradicting someone else in the company, only to have the customer cancel the job because we can't get our stories straight. Don't worry, your time will come, it's just not here yet. Your job right now is to assist the tech, listen, and learn all you can.

The purpose for all of this (aside from the obvious) is to give you, as the apprentice, more than just instructions, but a sense of responsibility, which will, in turn, breed pride. That pride will extend to your work in the house, and you may soon see some competition between install crews and their trucks. See if you can get your company to throw in a weekly $25.00 bonus for the cleanest truck, and then see what happens.

***

The Technician

On the jobsite the tech is responsible for cleanliness and quality of the work, finishing within the given time frame, safety, carrying out what is on the contract, and of course, making sure everything works. You need to be organized, be able to communicate well to both the customer and the helper, and be ready to delegate responsibilities. This actually sounds much simpler than it really is. Not because this is at all difficult to achieve, but because, like the helper, you also may never have been trained how to do your job properly. Most techs became techs because one summer, after working for a company for a couple of years, someone called in sick for a job that had to be done today, and there was nobody else to do it. So they gave the helper a shot, and with just a few minor blunders, here you are, the next mechanic. And that's fine, but somewhere in the equation should be some kind of formal training. After all, no one can be a marksman if he doesn't know how to line up the sights, or what the target is. Let's define the target.

Before any job begins, you should have been given a folder with all the necessary paperwork. This should contain things like contract, financing papers, job layout, etc. Don't skimp on reviewing all of these things in detail. Here you may find notes from the salesman, like, 'the customer must not hear any noise until after 10:00 AM' or 'watch out for the pet scorpions!' This may be your only chance to find out about all those idiosyncrasies of the job. Find out about the directions to the house, maybe it's not on the map, or there is no number on the mailbox. There's nothing worse than getting to the house all pissed off before you even start, just because of lousy directions. While you're taking care of these things, make sure your helper is cleaning and loading the truck, instead of talking about last night's ballgame.

When you get to the house, be sure both you and your crew are presentable and neat. Meet the customer, review the paperwork, get any necessary signatures and proceed to survey the house <u>with the customer</u>. Don't skip this step; this is your last chance to fill in any little gaps of understanding between the homeowner and your company before you start. Things like who is boxing in that exposed run in the bedroom, or "are you going to remove the whole house fan?" Clarify these things now. Next ask the customer for a staging area where you can store your stuff, and set up a work area.

At this point you should start putting a plan together of how you will execute the project. Will you work in the attic in the morning only, because it is too hot after that? Will you do all the ductwork first, because once the unit is in the attic, you won't be able to get to it? Or do you want to get the unit running first, and then work on the ductwork, so that you can at least cool the attic while you work? It is important that you figure this out now and devise a plan so that the job can run as quickly and as smoothly as possible, and with no confusion. While you are deciding on all of that, don't let the helper(s) stand there and watch you, direct them to start unloading and laying drop cloths. Remember, you're in charge.

Now gather your crew and layout the plan for them. The idea here is for you <u>not</u> to do all the work. Learn to let responsibilities go, piece by piece, as the crew can handle them. Watch what they do, and make sure it all gets done properly, but let them do it. I see way too many techs that let their helpers stand around and do nothing because they don't know how to do something. They want, and need to learn so they can build their confidence and advance their careers; but also so that you don't have to do it all yourself. Don't worry; they're not going to take your job away from you. Most techs are reluctant to teach for some reason, but really, a tech that can transfer his skills quickly and correctly to a helper, is extremely valuable to a company. It's not something to shy away from. This is a skill that companies seek out when looking for a new manager.

As each day comes to an end, contact your supervisor as they require, and give them a detailed progress report. They need to know about any problems that arise, issues with the customer, and any changes that may have happened during the day. Design conflicts are also a topic for discussion. What's on paper may not be practical, and before you change it, review it with whoever designed it. Many times there are things that need to be considered that you don't even know about.

Again, an important aspect of being a technician is teaching. You may never have thought of yourself this way, but as a lead tech, you may be the only training between the totally green helper, and his first installation on his own. Take the role seriously, it will make your job easier, more fun, and you will receive more respect from your peers. Not to mention that your manager is watching, and how your helper progresses will have a bearing on your pay and where you go from here. Someone, somewhere, took you under his wing, whether you realized it or not. Now it's your turn. Do it with pride!

Alright, what else does your job consist of? What about sales? Some companies don't want the field guys selling anything, but I think that's a waste of resources. In the end, we are all selling something to someone anyway. It may or may not be an actual product or service, but for now, let's talk about add-ons to the job at hand. Before your next install begins, talk to your manager about what sales opportunities exist for you with this customer along with what incentives may be available to you. Maybe the salesman forgot something like a special filter. Perhaps you can sell them a light and receptacle for the attic. Or maybe the customer just changed their mind, and decided to go with the 2-zone system after all. This could be very lucrative if a spiff or some type of commission is involved for you. Wouldn't you like to make a few hundred extra each month just for asking a few questions?

A media filter on each job could be enough to make the payments on a new boat! Believe it or not, it adds up quickly. There are many seminars out there, which address the whole sales procedure. They usually describe many closing techniques and ways to present your add-ons. A course like this can drastically increase your closing ratio. But before you have any closing ratio, you have to first ask the customers to buy. A very powerful method here is to bring into the house whatever it is you want them to buy, so they can see, and understand what it is that you're proposing. Then ask them if they would like you to add it into the rest of the job. Just by simply asking the question, some people will buy. Before you know it, you'll be on that new powerboat and off to trout-land. Remember, sales are everyone's responsibility.

CHAPTER 8
System Basics

When I sat down to write this chapter, I thought about it for awhile, then blew it off and went on to something else. The reason for that was, I didn't know how to contain it. I mean, we could talk about the 'basics' for the next 100 hours. What I finally decided I wanted to talk about here is simply a few straight forward system configurations, and types of equipment that you will probably have to work with at some point. Obviously, none of this is terribly technical, but at least the new guys will have some idea of what they are looking at. Much of this differs depending on what part of the country you work in, but this should cover most of you. Run through this quick so we can get back to the job at hand.

A/C ONLY SYSTEM

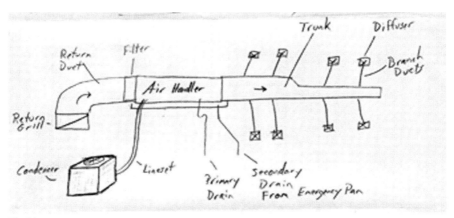

Fig 14

- Attic or basement
- Vertical or horizontal
- Unlike furnaces, most air handlers cannot accept a side return, so a stand must be fabricated for vertical application (as shown below)
- Basement system may require condensate pump

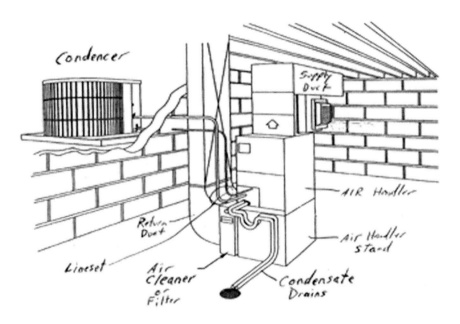

Fig 15

HEAT PUMP

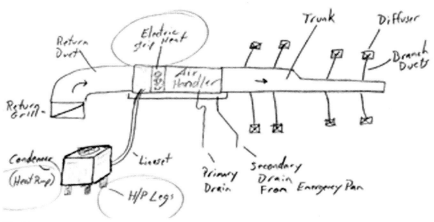

Fig 16

- Basically same as A/C with some additional items
- H/P condenser instead of A/C only. A heat pump is simply an air conditioner that has the ability to run backwards
- Condenser gets "legs" for operation in snowy climates
- Often has emergency or 2nd stage electric strip heat in air handler (requires additional electrical supply)
- Must use H/P thermostat

A/C SYSTEM WITH HYDROCOIL

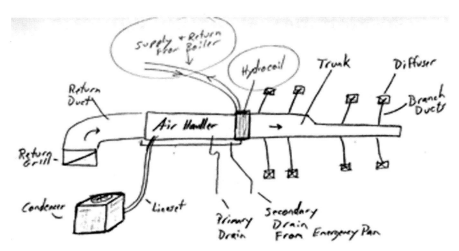

Fig 17

- A/C or H/P system with hydro coil built into air handler, or added on at end of unit
- Hot water, or steam supplied from remote boiler or water heater
- May require additional controls such as: circulator pump, zone valve, flow control, aqua-stat, etc.

GAS (NAT OR LP) FURNACE WITH A/C

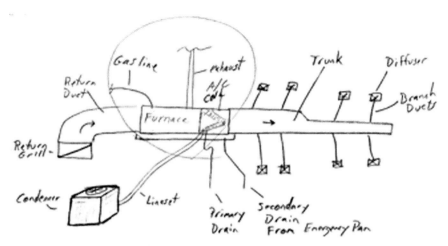

Fig 18

- 80% or 90% furnaces available
- 90% or condensing furnaces in attics may freeze (water in condensate line)
- Horizontal or vertical
- Use add-on A/C coil
- Requires addition of gas line
- 80% units require a chimney or power-venter
- 90% units usually vented with solid core PVC pipe
- Oil furnaces not used in attic
- Check local codes

DUCTLESS SPLITS

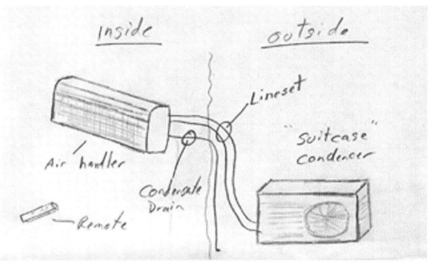

Fig 19

- Available in A/C or H/P
- Single or multiple air handlers from one condenser
- Used when ducted systems are not feasible
- Condensate line and line set often restrict air handler placement to outside wall
- Most manufacturers no longer offer heat pump with electric strip back-up
- Air handlers may be 24v, 120v, or 240v

(Do you think I over paid for the art lessons?)

PACKAGE UNITS

Fig 20

- Complete self-contained unit (condenser and air handler)
- Utilizes existing ductwork in house
- Most common on mobile homes
- Available heat pump models

HIGH VELOCITY SYSTEMS

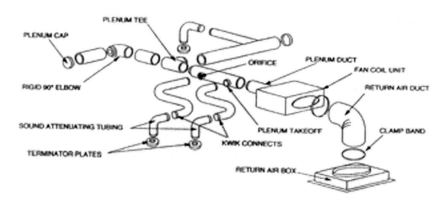

Fig 21

- Runs at much higher static pressure than traditional system
- Designed to fit into small areas
- Branch ductwork is 2" flex
- Must install minimum number of branch ducts (usually 5-7 per ton)
- Often twice the cost of conventional systems
- Accessories/options limited
- Hydro coils available
- Uses all proprietary ductwork

While there are many other systems out there, this small collection will comprise probably more than 90% of the jobs you do. Learn the basics of how these systems work, and the rest is just packaging. Hot water and steam boilers are common in the northeast and in some older cities which you may, or may not come across. Hydronic heating and commercial cooling systems are entirely different animals which we'll save for another time.

CHAPTER 9
Evaporator Coils & Air Handlers

The fact that you're reading this book, tells me that chances are good that this isn't your first exposure to the residential A/C installation (unless I somehow made it to the New York Times bestseller list). In fact, you probably think you know most of what there is to know about the subject. Maybe you've been doing this for 2 or 3 years now, and you've got it down. I'm telling you that it's the intelligent person who realizes that the longer you work in this, or any field, the more there is to learn. Just now, as you turned to this page, were you thinking that this is just first year air handler basics, and that you don't need to read it? Or were you open minded, thinking that you can always learn something? Well, why don't you keep reading for a little while longer, and I'll bet there is still something that you can learn.

In most residential, central air conditioning systems, there is an air handler (the indoor unit), and a separate condenser (the outdoor unit). This is commonly known as a split system. We'll get to the condenser in another chapter, but for now, let's talk about the air handler. The air handler usually can be found in an attic or a basement, or sometimes in both places. It consists of an evaporator coil, drain pan, some wiring, and a fan or blower. The cool refrigerant runs through the evaporator coil, the fan blows warm air from the house across it, and then the refrigerant absorbs the heat and takes it back to the condenser. When the warm air blows against the cold coils, moisture is removed from the air in the form of condensation, then drips into the drain pan. Sounds simple, but let's take a closer look.

Evaporator coils can be part of an air handler, or can be separate, for installation onto an existing furnace, in which case, the furnace already has a blower. Evaporator coils, or "coils", are usually "A" shaped, but can also be in the form of an "N" or an "I". The "I" coils are also known as "slab" coils. Each of these can be obtained in a case when used to sit on top of a furnace, or uncased to slip inside of the ductwork. While the BTU rating or size of the coil is very important, the actual, physical dimensions of the unit are also a concern. As with many components in this industry, just because you order a 3-ton coil, doesn't necessarily mean it will fit where the old 3-ton coil was. This is especially with the middle sizes, (2 ½ ton, 3 ton, 3 ½ ton). Most sizes come in not only different shapes, but also different widths and heights. For example, if you just order a 3-ton coil, you will need to specify vertical, horizontal, multi-positional, cased, uncased, 15", 17", or 20" width, not to mention r-22 or 410A, piston or TXV, plastic or metal drain pan, and the list continues. Did you know, for some reason, most companies don't make coils to match lowboy oil furnaces? I don't know why. Anyway, be specific and order the exact coil you need. Your job is tough enough not to. If you have to pan off the top of the furnace to match the coil, you'll look like an amateur for sure.

When placing a coil on top of a furnace, it is important to maintain a minimum distance between the top of the coil, and the top of the duct. Contrary to popular belief, the evaporator coil can extend into the trunk line as long as you follow these basic rules. The coil should not protrude into the horizontal section of the duct by more than one third, with the remaining space above it no less than the following: 4" above the coil on a 2 ton, 6" on a 2 ½–3 ½ ton, 9" on a 4 ton, and 12" on a 5 ton. These are good "rule of thumb" numbers to follow, without looking at each individual situation in detail.

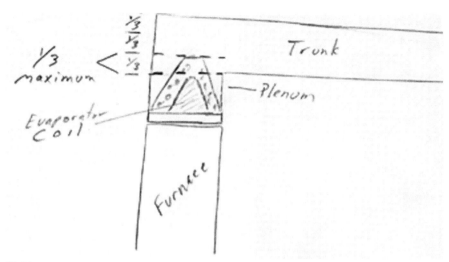

Fig 22

Aside from clearance on top of the evaporator coil, we need to be aware of what's happening underneath it also. If you are installing a coil on top of a furnace, whether it is oil or gas, you need to pay attention to one other detail. Most coils these days, cased or uncased, come equipped with plastic condensate pans, which usually work very nicely, and will not rot like the older metal pans. But have you ever gotten this phone call? Rrrrring. "Hello, Wally's Wonderful World of HVAC Service, this is Wally, can I help you?" "Yeah, hi. You guys installed a new A/C system on my furnace in May. It was great all summer, really worked well. But now that it's October, I turned on the heat, and smelled something burning from my furnace. Could you come over right away? Thanks." So, you run over there, take a look, and don't really see anything wrong, but definitely smell something odd. Luckily, you decide to investigate a little more, and think that maybe someone dropped a screwdriver or something into the furnace while doing the A/C job. It's possible. You go ahead, and take it apart. Uh Oh. Yup, you found the problem, all right. The drain pan from your new coil is melted all across the heat exchanger. You start to get that sinking feeling in your stomach, because you know that there is no quick, easy, or cheap fix out of this one. But *why* did this happen? Did the supply house give you the wrong coil? Go back and read the directions that came with the coil. You may find that the manufacturer <u>requires</u> a spacer between the furnace and the coil (usually 4") to avoid this type of mess. Once again, missing a couple of short sentences in the directions can cost the profit in the job, and more. This time, it's an evaporator coil, AND a new furnace. This is not a mistake you want to make twice.

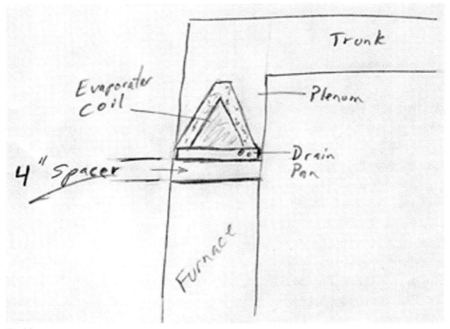

Fig 23

The typical air handler is very similar to the one shown in my ridiculously poor sketch below. Although usually one piece, some models are available in a modular configuration, that is, the evaporator coil comes separate from the blower section. These can be of particular advantage when you are working in an attic with trusses, or shallow roof pitch that was obviously never meant to have anyone up there taller than

3'2". Sure, the access hole in the ceiling is 30" x 30" and the unit is 22" x 26", so it will fit through the hole. But, will it make the turn? Some of you know what I'm talking about. (They all think it happens by magic).

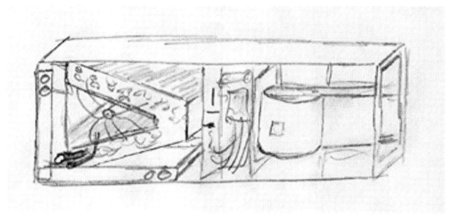

Fig 24

Most air handlers are multi-positional; or are they? Of the four possible positions, up flow, down flow, horizontal right and, horizontal left; most air handlers can convert to at least two. Be careful here. I'm sure many of you have unloaded all of your material at a jobsite and everything was going fine, only to find that the air handler you have can't be placed in the position you need. Naturally after it is in the attic. Now what? Send it back, and lose a half day. Do not pass go, do not collect a bonus.

Some air handlers are multi-positional, but do not come with a horizontal drain pan. Figure that out. Others require a special wiring adapter, to be ordered separately, if you are not adding electric strip heat. Which means that when it comes from the factory, you can't use the air handler at all unless you order the strip heat, or, order the no-heat kit? Huh? I don't know why either. The point is that you need to be familiar with whatever little quirks your brand of equipment has, so that you're not left out in the heat.

Speaking of heat, sometimes the A/C system you are installing will need to supply some form of heat to your customer, whether primary or secondary. This can be achieved in a number of ways, the two most common being, (1) electric strip, or, (2) a hydro coil, with hot water being supplied by some other source. Either of these can be added on to an air handler, or built into some brands. Changing the air handler to a furnace and coil, or the condenser to a heat pump are other options. BTU ratings for heat are separate, of course, from cooling ratings.

Yet another important item to keep an eye on in the replacement market is voltage. More than one installer has made the mistake of replacing an existing 110-volt air handler, with a 220-volt unit. That wouldn't be that big of a deal, if you found the line down at the electrical panel, and just changed the breaker to a double pole breaker, right? That is, until the installation is complete, and your helper starts to clean up, and you go to start the unit. POP!, Whoops!, @##$^%! What was that? Then you realize that the air handler wasn't on a dedicated line, it also had the attic fan, hall lights, and the bathroom on it. Now, not only do you have to go and run another wire from the panel to the air handler, but, you also have to go back and repair the damage. And what if there was no more space in that panel for that additional double-pole breaker, are you obligated to put in a sub panel, because you missed the voltage issue on the old unit? Poof, there goes that bonus again.

Seriously though, every time you have a problem, or make a mistake that the customer sees, you lessen your credibility with them. Yes, you can usually recover from a minor mishap, but many times the customer will begin to question every little thing you do. I'm sure you've all had the customer that inspects every inch of your work with a magnifying glass. These jobs never end. The two day job becomes four, because they want to be sure everything is okay, and you don't have a choice, because if you don't convince them that it's all good, you won't get paid. Be careful. Instill confidence.

I say all of this, mainly to stress the importance of the details before you start the job. Even if the salesperson missed this stuff, you can catch these things before you begin, and be the hero, rather than do a frustrating job, which ends up in a loss for your company.

All right, let's go back a few steps. Depending on what part of the country you live in, and if you've done a proper load calculation, you may find that it is pretty common for a house to call for a larger air handler, than condenser, such as a 3 ton air handler, and a 2 ½ ton condenser. Not only may this serve the house with proper dehumidification, but by over-sizing the evaporator coil as needed, you are also raising the overall SEER rating of the system (Seasonal Energy Efficiency Rating).

At this point, we need to review air handler ratings and sizes. As we discussed earlier, we need to look at both BTU's (preferably sensible), and CFM. Somewhere along the line, some intelligent soul decided to refer to both of these ratings in tons, however, they are definitely separate numbers. When choosing an air handler, or more specifically, an evaporator coil, you should check the manufacturer's specs regarding a particular unit. In general, when discussing BTU's (12,000 per ton) you can safely oversize the evaporator coil by a half a ton, but not the condenser. In other words, you can have a 3 ton condenser with a 3½ ton air handler (evaporator coil), but NOT vice versa. Installing the latter will produce an evaporator coil that will usually turn into a block of ice, giving you no airflow at all. With that in mind, let's take a look at two different manufacturers' air handler charts, and see what we find.

Performance data

AIRFLOW PERFORMANCE (CFM)

MODEL AND SIZE	BLOWER MOTOR SPEED	EXTERNAL STATIC PRESSURE (IN. WC)											
		0.10		0.20		0.30		0.40		0.50		0.60	
		208V	230V	208V	230V	208V	230V	208V	230V	208V	230V	208V	230V
FA4A 018	High	660	725	615	675	565	625	500	565	405	470	—	—
	Low	585	650	540	605	490	555	420	485	345	395	—	—
FB4A 018	High	660	695	815	870	765	820	715	760	645	690	550	600
	Medium	650	740	625	705	585	660	545	620	480	555	385	450
	Low	565	650	535	620	500	590	460	545	405	480	330	385
FA4A 024	High	940	975	890	925	835	865	780	805	715	735	635	650
	Low	820	900	785	855	745	805	700	750	640	680	545	575
FB4A, FC4B 024	High	945	975	900	930	840	870	780	805	695	725	560	595
	Medium	835	900	795	855	745	800	690	740	610	650	470	510
	Low	605	695	575	665	530	625	485	580	425	510	340	395
FA4A 030	High	1075	1170	1030	1115	985	1055	920	990	850	910	750	805
	Low	825	960	810	935	790	890	750	845	690	780	590	680
FB4A, FC4B 030	High	1260	1305	1290	1245	1135	1170	1065	1110	985	1015	880	900
	Medium	1055	1170	1020	1115	980	1055	930	1000	960	920	755	810
	Low	830	950	805	925	780	890	740	850	685	790	595	700
FA4A 036	High	1320	1405	1265	1345	1205	1280	1135	1210	1060	1120	960	1025
	Low	1100	1215	1070	1170	1020	1115	960	1060	890	980	805	895
FB4A, FC4B 036	High	1485	1580	1425	1490	1365	1420	1300	1350	1230	1275	1150	1190
	Medium	1235	1380	1200	1325	1160	1265	1110	1210	1055	1140	985	1070
	Low	1035	1185	1010	1150	980	1115	940	1070	890	1010	825	935
FA4A, FB4A, FC4B 042	High	1580	1710	1540	1655	1495	1595	1440	1530	1375	1445	1290	1355
	Medium	1400	1570	1375	1525	1350	1480	1305	1425	1255	1360	1175	1280
	Low	1195	1375	1180	1350	1165	1325	113?	1285	1085	1240	1020	1160
FA4A, FB4A, FC4B 0??	Hi		935	1785	1830	1700	?5			1520	1555	1430	
			1660							143?		13?	

Fig 25

One of the first things you may notice is that a three-ton air handler doesn't just put out 1200 CFM (400 CFM per ton). There are a number of variables that affect how much air is realized from a unit. On the first chart there is static pressure across the top. For most residential systems, you will usually design a supply duct system at 0.1 W.C. Next, you will see that each size has a low, medium and high speed. These speeds are adjustable at the air handler only, not by the homeowner. Some air handlers have only 2 speeds, some have 4. The 036 (3 ton) which you think of as producing 1200 CFM, now is shown to throw as little as 1035 on low, and as much as 1485 on high. Go one step farther and look at how much the ratings change with the differences in voltage. So, from one extreme to the next, with the same unit, you can get from 1035 CFM to 1520 CFM. These factors are often overlooked, but can shed an awful lot of light on some of your current problem systems. When you go back to try and solve some of those problems, you may be surprised at what you find. It's happened to me more than once. From one manufacturer to another, ratings can change dramatically. Let's look at another.

UNIT MODEL	MOTOR HP-AMPS (120V)	MOTOR SPEED CONN.	CFM vs. EXTERNAL STATIC PRESSURE					
			0.05	0.10	0.20	0.30	0.40	0.50
18MBXB-HW	1/5 - 2.8	HIGH	810	780	715	650	580	500
		MED.	680	655	600	545	490	420
		MED. LOW	530	505	460	415	360	295
		LOW	350	325	270	220	160	- - -
24MBXB-HW	1/5 - 5.1	HIGH	950	920	855	790	720	645
		MED.	860	835	785	720	650	580
		LOW	780	755	705	650	590	510
30MBXB-HW	1/5 - 5.1	HIGH	1120	1095	1045	995	940	880
		MED.	850	840	810	780	740	690
		LOW	680	670	655	625	585	510
36MBXB-HW	1/2 - 8.5	HIGH	1340	1310	1250	1190	1120	1050
		MED.	1290	1260	1200	1140	1080	1000
		LOW	1200	1170	1120	1070	1010	940
48MBXB-HW	3/4 - 10.7	HIGH	1810	1780	1720	1660	1590	1530
		MED.	1570	1550	1510	1460	1400	1340
		LOW	1280	1260	1220	1180	1130	1050
60MBXB-HW	1 - 11.5	HIGH	2160	2125	2055	1980	1895	1810
		MED.	1865	1840	1785	1710	1620	1525
		LOW	1560	1540	1490	1435	1365	1260

Fig 26

If you look at this same size unit from a different company, the differences are apparent. Look at the CFM produced here, compared to the first unit's chart. Manufacturer's charts vary greatly with the amount of info they supply you with. The better and more expensive units generally place a higher value on the engineering and design info they give you. The chart below, taken from a Carrier condenser manual, also gives you a SEER for each combination of air handler, and evaporator coils, matched with certain condensers. As you can see, the manufacturer acknowledges the need for different size unit combinations. You will again notice how the SEER can be raised for a complete system without changing to a more efficient condenser. (Yes, I know it's an old chart) Use this against your competition if you are selling the system yourself. The other company probably wouldn't even know what you're talking about. There you go again, building credibility with the customer.

Another thing to discuss in this chapter is the variable speed air handler. This is a more involved and sophisticated unit, which will not only raise the SEER significantly, but will also create the most comfortable air system available. The air handler itself will decide how fast its fan should run to not just cool the house, but more importantly, maintain a very tight temperature range. It's much quieter, more efficient, and will filter the air much more effectively. Once you experience life with a variable speed blower, you will probably never go back to anything else. If this is something that you will be installing in your company, I suggest that you get in touch with your local distributor and find out when they are having their next variable speed training class. They may even be willing to come to you. Take advantage of it, it's well worth the time.

Combination ratings continued

UNIT -SERIES	INDOOR MODEL	TOT. CAP. BTUH	FACTORY-SUPPLIED ENHANCEMENT	STANDARD RATING	CARRIER GAS FURNACE OR ACCESSORY TDR†	ACCESSORY TXV‡	SOUND RATING (dBA)
	CK3BA036	33,400	TDR	10.50	—	10.50	82
	CK3BA042	33,400	TDR	10.80	—	10.80	82
	COILS + 58MVP120-20 VARIABLE-SPEED FURNACE						
	CJ5A/CK5A/CK5BA042	33,400	TDR	10.80	—	10.80	82
	CJ5A/CK5A/CK5BW036	33,400	TDR	10.50	—	10.50	82
	CK3BA036	33,400	TDR	10.50	—	10.50	82
	CK3BA042	33,400	TDR	10.80	—	10.80	82
	COILS + 58U(H,X)V060-12 VARIABLE-SPEED FURNACE						
	CC5A/CD5A/CD5BA036	33,400	TDR	10.50	—	10.50	82
	CC5A/CD5A/CD5BA042	33,400	TDR	11.00	—	11.00	82
	CC5A/CD5A/CD5BW042	33,400	TDR	11.00	—	11.00	82
	CD3(A,B)A036	33,400	TDR	10.50	—	10.50	82
	CD3(A,B)A042	33,400	TDR	11.00	—	11.00	82
	CD3CA036	32,400	TDR	10.50	—	10.50	82
	CD3CA042	32,600	TDR	10.50	—	10.50	82
	CD5A/CD5BW036	33,400	TDR	10.50	—	10.50	82
	CE3AA036	33,400	TDR	10.50	—	10.50	82
	CE3AA042	33,400	TDR	11.00	—	11.00	82
	CJ5A/CK5A/CK5BA036	33,400	TDR	10.50	. —	10.50	82
	CJ5A/CK5A/CK5BN036	31,400	TDR	10.50	—	10.50	82
	CK3BA036	33,400	TDR	10.50	—	10.50	82
	CK3BA042	33,400	TDR	10.80	—	10.80	82
	COILS + 58U(H,X)V080-16 VARIABLE-SPEED FURNACE						
	CC5A/CD5A/CD5BA036	33,400	TDR	10.50	—	10.50	82
	CC5A/CD5A/CD5BA042	33,400	TDR	11.00	—	11.00	82
	CC5A/CD5A/CD5BW042	33,400	TDR	11.00	—	11.00	82
	CD3(A,B)A036	33,400	TDR	10.50	—	10.50	82
	CD3(A,B)A042	33,400	TDR	11.00	—	11.00	82
	CD3CA036	32,400	TDR	10.50	—	10.50	82
	CD3CA042	32,600	TDR	10.50	—	10.50	82
	CD5A/CD5BW036	33,400	TDR	10.50	—	10.50	82
	CE3AA036	33,400	TDR	10.50	—	10.50	82
36-30, 31, 32, 33, 50, 52, 62	CE3AA042	33,400	TDR	11.00	—	11.00	82
	CJ5A/CK5A/CK5BA042	33,400	TDR	10.80	—	10.80	82
	CJ5A/CK5A/CK5BW036	33,400	TDR	10.50	—	10.50	82
	CK3BA036	33,400	TDR	10.50	—	10.50	82
	CK3BA042	33,400	TDR	10.80	—	10.80	82
	COILS + 58U(H,X)V100-20 VARIABLE-SPEED FURNACE						
	CC5A/CD5A/CD5BA036	33,400	TDR	10.50	—	10.50	82
	CC5A/CD5A/CD5BA042	33,400	TDR	11.00	—	11.00	82
	CC5A/CD5A/CD5BW042	33,400	TDR	11.00	—	11.00	82
	CD3(A,B)A036	33,400	TDR	10.50	—	10.50	82
	CD3(A,B)A042	33,400	TDR	11.00	—	11.00	82
	CD3CA036	32,400	TDR	10.50	—	10.50	82
	CD3CA042	32,600	TDR	10.50	—	10.50	82
	CD5A/CD5BW036	33,400	TDR	10.50	—	10.50	82
	CE3AA036	33,400	TDR	10.50	—	10.50	82
	CE3AA042	33,400	TDR	11.00	—	11.00	82
	CJ5A/CK5A/CK5BA042	33,400	TDR	10.80	—	10.80	82
	CJ5A/CK5A/CK5BW036	33,400	TDR	10.50	—	10.50	82
	CK3BA036	33,400	TDR	10.50	—	10.50	82
	CK3BA042	33,400	TDR	10.80	—	10.80	82
	COILS + 58U(H,X)V120-20 VARIABLE-SPEED FURNACE						
	CC5A/CD5A/CD5BA036	33,400	TDR	10.50	—	10.50	82
	CC5A/CD5A/CD5BA042	33,400	TDR	11.00	—	11.00	82
	CC5A/CD5A/CD5BW042	33,400	TDR	11.00	—	11.00	82
	CD3(A,B)A036	33,400	TDR	10.50	—	10.50	82
	CD3(A,B)A042	33,400	TDR	11.00	—	11.00	82
	CD3CA036	32,400	TDR	10.50	—	10.50	82
	CD3CA042	32,600	TDR	10.50	—	10.50	82
	CD5A/CD5BW036	33,400	TDR	10.50	—	10.50	82
	CE3AA036	33,400	TDR	10.50	—	10.50	82
	CE3AA042	33,400	TDR	11.00	—	11.00	82
	CJ5A/CK5A/CK5BA042	33,400	TDR	10.80	—	10.80	82
	CJ5A/CK5A/CK5BW036	33,400	TDR	10.50	—	10.50	82
	CK3BA036	33,400	TDR	10.50	—	10.50	82
	CK3BA042	33,400	TDR	10.80	—	10.80	82
	CC5A/CD5A/CD5BA042*	40,000	NONE	10.00	10.10	10.10	82
	CC5A/CD5A/CD5BC048	39,500	NONE	10.00	10.10	10.10	82
	CC5A/CD5A/CD5BW042	40,000	NONE	10.00	10.10	10.10	82
	CC5A/CD5A/CD5BW048	40,000	NONE	10.00	10.10	10.10	82
	CD3(A,B)A042	40,000	NONE	10.00	10.10	10.10	82
	CD3(A,B)A048	40,000	NONE	10.00	10.10	10.10	82
	CD3CA042	38,000	NONE	—	10.00	10.00	82
042-30, 31, 50, 51, 61	CD5A/CD5BA048	40,000	NONE	10.00	10.10	10.10	82
	CE3AA042	39,500	NONE	10.00	10.10	10.10	82
	CE3AA048	40,500	NONE	10.00	10.10	10.10	82
	CG5AA048	40,500	TXV	10.00	10.10	—	82
	CJ5A/CK5A/CK5BA042	40,000	NONE	10.00	10.10	10.10	82
	CJ5A/CK5A/CK5BA048	40,000	NONE	10.00	10.10	10.10	82
	CJ5A/CK5A/CK5BN042	39,000	NONE	10.00	10.10	10.10	82
	CJ5A/CK5A/CK5BN048	39,000	NONE	10.00	10.10	10.10	82

Fig 27

The last thing in this section is often overlooked, but is at least as important as anything else here, and that is return air. Without enough return air, you can throw out most of what we just talked about. Probably a full three quarters of all air conditioning problems have to do with a lack of return air. Some of these we will discuss in a later chapter, but why the subject doesn't get more attention is beyond me. Think about this; if your return ductwork is only allowing the air handler to bring in 900 CFM, how do you expect it to deliver 1200 CFM on the supply side of a three ton unit? Magic? Without enough return air, all the other calculations and planning you've done, mean about as much as a pair of sunglasses in a nuclear blast. And remember that return air pressures should be much lower than on the supply side, which means larger ductwork. Don't say it doesn't matter, it does. Lack of proper return air is the cause of many callbacks, such as noisy systems and condensate leaks. Don't let the return suck up the profit.

CHAPTER 10
The Condenser And The Line Set

The big, heavy thing that goes outside, next to the house, is called the condenser. Its purpose is to cool the gas, or more accurately, to remove the heat from the refrigerant. It does this, basically, by first running through an outside coil, then through a compressor, which, hang on to your seats, kids, compresses the gas back into a liquid. That completes our physics lesson for today. Refrigeration theory deserves a dedicated course and book which I hope you will place a high priority on, and refer to as needed.

The condenser, staying true to form with other A/C components, is sized in tons. Common residential sizes include 1½, 2, 2½, 3, 3½, 4 and 5-ton units (there is no 4 ½ ton). Most manufacturers identify the unit's size in the model number by a multiple of 12. In other words, a 2-ton unit would have a 24 in the number, and a 3½-ton would have a 42, such as: 38TKB0**42**010. SEER is the next rating condensers are usually noted by. This is the yellow sticker that the government requires, to identify the efficiency rating, as in Seasonal Energy Efficiency Rating. I like to refer to it as a 'miles per gallon' type number. That is, the higher the number, the more efficient the unit. Currently, 13 is the minimum, and can go as high as 22 SEER. As SEER ratings increase, typically, so does the physical size of the unit. The larger the surface area of the condenser coil, the greater the ability to dissipate the heat, which brings us right into heat pumps.

Efficiency is even more important when used in context with a heat pump, because it is likely that a heat pump will see much more use than an A/C only system. Here, you will find EER ratings. With the increased use of a heat pump, you will probably also see a slightly shorter life of the unit.

A heat pump is, very simply, an air conditioning system that has the ability to run backwards. In heating mode, it absorbs heat through the refrigerant from outside, and brings it in to heat the house. All of the systems we are talking about in this book are atmospheric systems. A Geothermal heat pump is another type of system that uses a source of ground water in an open or closed loop to absorb or dissipate heat. You may also come across water-cooled chillers for air conditioning, however, they are usually only found in commercial systems. Both geothermal and chillers are left out of this project.

The heart of the condenser is the compressor. There are only a handful of compressor manufacturers in the United States, and most are not the condenser manufacturers themselves. Scroll compressors are all the rage today, with most customers learning about them on the Internet. A scroll compressor, as you may already know, has only two moving parts compared to a reciprocating compressor, which has a few more. As far as compressors go, you should get familiar with what goes on in them, however if you come across a bad one, they are not repairable, only replaceable.

Another rating you may find on condenser literature is a decibel rating, usually somewhere between 78 and 83. This relates to how much noise a unit makes. This will be a consideration to the homeowner if they have a small yard, or maybe the unit will be near the deck or picnic table. Manufacturers use a number of different methods to quiet their condensers, including the use of a sound hood, (which aerodynamically passes air through the unit with less friction), blankets of insulation over the compressor, and even layers of rubber between metal surfaces to deaden the tinny sounds.

Noise is an item you can use to upgrade your sale. While efficiency may not be important to a particular customer, because their unit has to go next the master bedroom window, sound may be very important. In general, the higher efficiency units (read more expensive) are usually quieter. Obviously, this is a conversation to have *before* the unit is installed.

Combination ratings continued

UNIT E-SERIES	INDOOR MODEL	TOT. CAP. BTUH	FACTORY-SUPPLIED ENHANCE-MENT	STANDARD RATING	CARRIER GAS FURNACE OR ACCESSORY TDR†	ACCESSORY TXV‡	SOUND RATING (dBA)
	CK3BA036	33,400	TDR	10.50	—	10.50	82
	CK3BA042	33,400	TDR	10.80	—	10.80	82
COILS + 58MVP120-20 VARIABLE-SPEED FURNACE							
	CJ5A/CK5A/CK5BA042	33,400	TDR	10.80	—	10.80	82
	CJ5A/CK5A/CK5BW036	33,400	TDR	10.50	—	10.50	82
	CK3BA036	33,400	TDR	10.50	—	10.50	82
	CK3BA042	33,400	TDR	10.80	—	10.80	82
COILS + 58U(H,X)V060-12 VARIABLE-SPEED FURNACE							
	CC5A/CD5A/CD5BA036	33,400	TDR	10.50	—	10.50	82
	CC5A/CD5A/CD5BA042	33,400	TDR	11.00	—	11.00	82
	CC5A/CD5A/CD5BW042	33,400	TDR	11.00	—	11.00	82
	CD3(A,B)A036	33,400	TDR	10.50	—	10.50	82
	CD3(A,B)A042	33,400	TDR	11.00	—	11.00	82
	CD3CA036	32,400	TDR	10.50	—	10.50	82
	CD3CA042	32,600	TDR	10.50	—	10.50	82
	CD5A/CD5BW036	33,400	TDR	10.50	—	10.50	82
	CE3AA036	33,400	TDR	10.50	—	10.50	82
	CE3AA042	33,400	TDR	11.00	—	11.00	82
	CJ5A/CK5A/CK5BA036	33,400	TDR	10.50	—	10.50	82
	CJ5A/CK5A/CK5BN036	31,400	TDR	10.50	—	10.50	82
	CK3BA036	33,400	TDR	10.50	—	10.50	82
	CK3BA042	33,400	TDR	10.80	—	10.80	82
COILS + 58U(H,X)V080-16 VARIABLE-SPEED FURNACE							
	CC5A/CD5A/CD5BA036	33,400	TDR	10.50	—	10.50	82
	CC5A/CD5A/CD5BA042	33,400	TDR	11.00	—	11.00	82
	CC5A/CD5A/CD5BW042	33,400	TDR	11.00	—	11.00	82
	CD3(A,B)A036	33,400	TDR	10.50	—	10.50	82
	CD3(A,B)A042	33,400	TDR	11.00	—	11.00	82
	CD3CA036	32,400	TDR	10.50	—	10.50	82
	CD3CA042	32,600	TDR	10.50	—	10.50	82
	CD5A/CD5BW036	3?400	TDR	10.50	—	10.50	82
	CE3AA036		TDR	10.50	—	10.50	82
036-30, 31, 32, 33, 50, 52, 62	CE3AA042		TDR	11.00	—	11.00	82
	CJ5A/CK5A/CK5BA0??		TDR	10.80	—	90	82
	CJ5A/CK5A/CK???	?DR	10.50				82
	CK3BA??	??		10.50			82
	C???			10.80			82
VARIABLE							

Fig 28

Speaking of placement of the condenser, many customers ask about putting the unit somewhere out of the sun. I have never seen a situation where having the unit in direct sunlight has made any noticeable difference in cooling performance. What I have seen though, are customers, or even an installer, who was obsessed with the sun issue enough to cover the unit with bushes or some sort of fencing. If you don't already know, the four sides of the condenser all have minimum clearances by the manufacturer specifically to eliminate these types of obstructions. The condenser needs to have plenty of air across its coils to expel all of the heat it took out of the house. Along with side clearances, there is also a minimum top clearance. Installers sometimes go overboard with this one. Most manufacturers recommend somewhere between 3 and 5 feet of clearance above the unit. Twenty-six feet is not necessary. It is perfectly acceptable, in most cases, to put the condenser under a deck, if the deck is six or seven feet high, has open sides, and a slotted floor.

There are other clearance considerations, however. When choosing a location for a new condenser, you should first try to find an area around the house that has less traffic, mostly for noise consideration, as already mentioned, along with the side and top clearances stated by the manufacturer. You'll need to level the ground under the unit and should place it on a pad of sorts, concrete or fiberglass is common. The next consideration has to do with local codes. Codes regarding fuel tanks and gas regulators need to be followed. Propane tanks usually need a clearance of ten feet from any electrical appliance, such as a condenser. I once had a militant propane delivery guy report me to his company because we were only 9½ feet from their tank. I don't even know what to say about that. Anyway, it's also a good idea to talk to your local utility company to see what distances they require from the gas regulator, or meter; they usually only need about 3 feet.

In the last few chapters, we've talked generically about "gas" or refrigerant. Let's look at some names. R-22 (more commonly known as Freon) has been the most common refrigerant used in residential A/C systems until the last few years. As the world becomes more ecologically aware, chlorine based refrigerants have become as popular as lice, due to their destructive nature to the atmosphere. Now R-410A has emerged as the new standard. The government requires that anyone who works with these gases to be EPA certified. Any local supplier should be able to set you up with a class to get certified. Learn and use these procedures.

There are big fines if you get caught releasing refrigerant out into the atmosphere, and it's just not an environmentally sound thing to do. The manufacture of R-22 is about to be curtailed, and equipment using R-22 should no longer be manufactured as of 2010. As I said, the new refrigerant standard for residential systems is R-410A, (originally known as Puron), which was developed by Carrier and DuPont. The 410A containers are pink instead of green (R-22). Keep in mind that this is _not_ a direct replacement for R-22. R-410A runs at much higher pressures, with the high side around 400lbs instead of 250lbs, and will _not_ be compatible with most of the existing equipment out there. Among other reasons, R-22 equipment is not designed to run anywhere near that kind of pressure. Oh, and your current gauges won't work either.

Alright boys and girls, let's move on. To connect the air handler to the condenser, we use two copper lines, called the line set. One is for liquid, and the other is for gas or suction. (Basic refrigeration is a course that all technicians should take as soon as they enter this industry. The theory and workings of refrigeration warrant a stand-alone class, which is why we won't get into it in this book.) Most residential line sets come in a roll, with the suction line pre insulated. The first item of business here is to unroll it. Not so fast, there's a technique to this.

Every one of us who has had a new helper or friend (or ourselves) on an A/C job wants to unroll the line set. Why, I don't know, looks like fun, maybe? So they jump right in, begin unrolling, and end up with 5 kinks in it before they get half way done. This is why you want to stand it up, hold the piping to the ground with your foot, and unroll four feet at a time, then move your foot closer to the roll again. And, don't remove the plugs until you're ready to braze. Once you have it rolled out on the lawn, now is the time to tape the low voltage wire to it so you can eliminate that step later. Then, install the line set where you planned, taking care not to rip the foam, nor kink the copper. Speaking of the foam insulation, if it does get ripped or torn, be sure to fix it well because any area of the suction line that is exposed to the air will collect condensation and will drip, a lot. Once it's straight, secure it, and braze it without setting the house on fire. Don't laugh. When brazing a refrigerant line, get in the practice of running nitrogen through it which will prevent yucky stuff from forming inside as you braze.

When the connections are made, the system is then evacuated (meaning, we remove all of the air and moisture by creating a vacuum) with a pump, and should then be cleaned out with nitrogen. Impurities in refrigerant are compressor killers, so make sure you follow the manufacturer's recommendations. A filter drier to catch these impurities, while not necessary on a new installation, will certainly help the system last longer. Some brands come with small filter-driers from the factory. Line sets, incidentally, come with liquid lines usually in 3/8" copper tubing, and suction lines in 5/8, ¾, 7/8, and now, 1 1/8" is common. They come in pre-cut, sealed, and insulated lengths, such as

25', 35' and 50'. The cleaned, sealed, and evacuated tubes are evidenced by the rubber plugs at the ends, and shouldn't be removed until you are ready to braze them in.

The new condenser comes from the factory, pre-charged with the appropriate gas. How much gas is a good question. Most units come pre- charged with 15 feet of refrigerant. You still need to look at the installation manual for each condenser to find out exactly how much, and it will also explain how much refrigerant to add per foot of line set. By knowing this number, such as .75 ounces per foot, you are then able to weigh in the correct charge to the system using a digital scale without guessing. The manual will also tell you how long of a line set you can use, and all the particulars that the manufacturer would have you follow.

Let's discuss line sets in a bit more detail because they're so exciting. You know, like staplers or fog. If you are installing a new system, you will need to scope out a path for the piping. First, when it comes to refrigerant piping, shorter is better. In most situations, the path from the coil to the condenser will be obvious. Sometimes, though, it won't be. Hopefully someone planned this out before the job started and didn't take it for granted. If the coil will be near an outside wall, you'll probably be O.K. But how will you run it if it goes in a closet in the middle of a finished basement? Can you sneak it under the sheetrock between the joists? What if the joists are going the other direction? If it is going to be covered in a wall or ceiling, do everything you can to do it in one piece without any joints (I mean the pipe, err, the refrigerant pipe). You should also know that these lines can vibrate in the wall or ceiling, so remember to secure them properly.

There will definitely be times when you need to be very creative as far as where you run the line set, and how you cover it. You may be doing an attic system and have to cross another roof outside, from an addition. That's NOT easy to camouflage and you should look for alternatives, like running it through a knee wall or closet to get to where you need to go. Out of about 3000 installations that I've been involved in, I can only think of a handful that we couldn't resolve satisfactorily. When you do get the piping outside to run down the side of the house, you'll need to cover it with something. Leader piping that matches the gutters of the house usually works well. It's light, has different elbows and fittings available, and can be cut near the seam with a utility knife so you can place it around the piping instead of trying to thread it through. You can also find white, 2 piece pipe cover for this exact application. With this stuff, you mount the back piece to the wall, place the line set in it, and then just snap on the cover.

One last situation you should give some thought to. There are an awful lot of condos and townhouses out there that were built with air conditioning. For some reason manufacturers must think of these as their laboratories, because this is where you find all of those oddball systems that they only made for two months, before they decided "that doesn't work so well". Aside from that, most of those systems live in a closet in the middle of the house, with the line set buried in a wall or ceiling that runs right through the living room, or even in the concrete slab. Yes, they really did. Worse yet, those are not only all R-22 systems, but many are old enough to have ¼ x 5/8 line sets, too small for the new system, even if you did take a chance on just cleaning it out. Wait until you get one that came with a pre-charged, corrugated line set covered with rubber that you can't even adapt too. Yeah! Just be aware that it's not always across the ceiling, and pop through the wall. More than once I had no other option than to set the condenser away from house, and run the line set underground in a PVC conduit. Be creative, you'll figure it out. Now, back to condensers.

In a previous chapter, we talked about how when doing a proper load calculation, you will often find that a certain house may call for a larger air handler than condenser, which is fine. Don't forget to put the matching orifice that comes with the condenser into the evaporator coil, or you will negate any of the benefits of the larger unit. These days, most air handlers come with a thermal expansion valve (TXV) instead, so you don't need to worry about orifices and pistons as much. Again, check with the manufacturer's instructions. The TXV, by the way, is the more accurate of the two and will result in a more efficient system.

When brazing the line set to the condenser, remember to remove the Schrader pins, braze with nitrogen in the lines to prevent impurities, and also be careful not to overheat the service valves. Some manufacturer's valves are more sensitive to heat than others, and are easily damaged by a torch, hence requiring replacement. By the laws of nature, this only happens on the last day of the job, 45 minutes before you have to be home for your daughter's soccer game. So defy your instincts and take an extra 18 seconds to place a wet rag around them.

CHAPTER 11
Filters, Accessories, And Other Add-ons

Aside from the equipment and duct system improvements that the sales guys talk about, there are countless other components and upgrade that are available and waiting to be bought by the homeowner. These extras can dramatically improve not only the efficiency, comfort, and usability of the systems, but more importantly, your bank account. Let's take a virtual stroll (in my head) around a typical house and see what we can find.

The name of this game is observation and imagination. Here's how it works. While you are working in your customer's home, you pay attention to what the household consists of, such as people, cars, hobbies, etc. Now, don't act like you're casing the joint, just notice what's going on. Next imagine how you can improve this family's life. While you are putting together duct work, or pulling wire, begin to make a list of what you think these people may need to make life easier.

Is it pollen season? Pollen is great stuff. It's the smell of another bonus. Don't be shy, ask them if anyone in the house has allergies. Probably one out of every two homes you are in will have someone who suffers from them. Being the dutiful installer that you are, inform your host of the different possibilities that exist in order to relieve them of their pain.

They'll probably say something like. "Doesn't it come with a filter?"

Then You: "Why, yes it does come with a 1" fibrous filter that is really meant to protect the equipment, more than anything else."

Them: "Won't that remove the pollen in the air?"

You: "Actually, no. As far as air quality goes, those filters are good at catching socks and small animals, but that's about it."

Them: "Oh my, that just won't do! What do you recommend? Please help us!"

You: "Well, nice customer let me give you a few choices. The first upgrade would be the electrostatic filter. This will fit in place of your other filter, and is a permanent, washable filter, which works by self generated static electricity to attract dust particles."

Them: "Oooh, ahhh, etc. etc."

You: "Then, from there we go up to the media filter. This is a resistance type filter, which looks like a pleated cloth filter. These are very effective, and require only minimal maintenance. Simply change the cartridge every six to twelve months, and you're set. Best of all, people tell us that they don't have to dust as much anymore."

Them: (stunned and amazed) "Wow!"

You: "Finally, there is the electronic filter. Very effective, but a bit more maintenance, and fairly costly. What I would suggest is, a U.V. light which will kill the mold and bacteria in the air, along with the media filter I described. With those, your lives will be perfect!"

Them: "Heavens to Mergatroid! How did we ever get by without you? Can we buy all of them? Please?"

See, it's that simple!

All right, back to business. Filters are great upgrades, and you can even go further if necessary. Indoor air quality is definitely a hot topic these days, and all of these items fit the bill. Most homes could greatly benefit from a humidifier (for winter use, of course). Did you know that with the correct amount of moisture in the home your customer will feel more comfortable at a lower temperature? We're talking fuel savings here.

While we're talking about things that go inside a duct system, let's just mention that there are smoke detectors available, which are usually required in commercial systems, but can be installed in homes as well.

Most of these items discussed so far in this section, require little or no extra work, and should not extend the installation time.

Another component that is becoming more common, especially on very tight, or energy efficient homes, is the Heat or Energy Recovery Ventilators. This is a heat exchanger that allows outside air to enter the home, but first heats or cools it with the air that it is replacing. They are very efficient, and customers are going to ask you about them, so you should know something about them. Again these are easy to install, do just one, and you're an expert.

Thermostats, you say? Yes, there are probably as many thermostats as you have tools in your truck. Seven day programmable, heat only, cool and heat, 2-stage H&C with backup, large number, turn the dial, line voltage, landlord, remote sensing, network, wireless, whatever it is that you want, you'll have to choose between more than you wanted to know about. Get familiar with three or four of them, and leave it at that. Don't go crazy here, but there are a lot of really cool systems out there that the hi-tech guys will go crazy with. These are great add-ons to what you are already putting in.

Accessories and add-ons can go on forever. Whatever you see a need for is a potential add-on, and some bonus money. Pull down stairs, an attic light with a switch, or maybe a receptacle, which would make yours, as well as the homeowner's and the next service guy's, trip up to the attic all a little easier. What about a condenser cover, new or upgraded diffusers (brass are nice) or how about another...ZONE?

Ah, the 2nd zone! If you know anything about them, it's probably stirring a few emotions right about now. There are some complicated systems out there that do all sorts of impressive things. There are also very simple systems that don't really do much of anything, past the basics. Whatever it is that your company sells, in order to install it properly, you need a basic understanding of how it works. Zoning is a great upgrade, but again, make sure you know how to do it right. When most thermostats call for cooling, the air handler and condenser will react by blowing cold air to that zone. Keep in mind, that air handler is going to put out its full capacity, whether one zone damper is open, or all four are open. Now think about that, if only one zone calls, doesn't that mean that all of the air required to cool the whole house is going to just one zone? Yes it does! So if you don't plan in some type of waste gate (like in the world of turbochargers), or barometric damper, that one zone is going to be one noisy wind tunnel. This is a serious comfort issue, so make sure you address this. All that excess air has to go somewhere. Find a dump zone.

When a salesperson goes out and sells an air conditioning system, many times they sell with price as the biggest concern, and they don't consider all the performance factors. Cooling two floors of a house on one single zone is usually okay. You can build in the balance for, let's say, a colonial with an attic system in one zone. But add heat to that same system in the same house, and it will NEVER work on one zone. I'm going to tell you why, keep your shirt on. In the summertime, when it's 95° outside, where is the hottest place in that colonial? In the basement? No, it's on the second floor, right? So that's where we'll put all the air, upstairs, so now it will be nice and cool, and everyone will be happy. Now in January, when it is 15° below zero outside, what is the coldest part of that home? The bedrooms? Nope, it's next to the front door, and in the kitchen by the sliding door. So that's where we'll put the heat, downstairs. But wait, you already put most of the ductwork upstairs for the A/C! Get the point? Heating and cooling on _one_ zone, on _one_ floor, is fine. Cooling _or_ heating on two floors works okay too, but _not_ heating _and_ cooling on two floors with one zone. It won't work, so don't even try it.

Zoning is an excellent add-on, which can solve many problems, and you can make a lot of money with it. If that is going to be something you want to focus on, do it right, take a course, read some books, and best of all, talk to people that do a lot of it. You can also speak to the manufacturers' representatives in your area. They are always looking for people to spill all their knowledge on. They will usually be more than happy to come to your shop and teach you all about their products, and how to install them properly. You can probably even get them to come to your jobsite. These days, while many of the systems can do more and more, they sometimes actually do get easier. Some companies are now selling wireless thermostats, which will save you some time, and also make you 'high-tech' compared to your competition.

An add-on, is an add-on. (How's that for a little philosophical insight?) Just because you're putting in a new furnace, or air conditioning system, doesn't mean you can't sell a new hose bib or outdoor faucet to that homeowner. Think about it, how many people do you know that wouldn't like to have a hose bib on the other side of the house, by the garden or driveway? I'll bet you could sell five or ten a month if you tried. Oh, by the way, you don't even need any selling skills here. I guarantee that if you just *suggest* one add-on at each job, you will substantially increase your pay.

Here are some ideas, try 'em and watch your bank account fill up:

Upgraded filters	Add air conditioning to furnace
Humidifiers	Heat/cool the garage
Zone systems	Heat/cool the basement
Thermostats	Laundry sinks
Condenser covers	Heat Recovery Ventilation unit
Different diffusers	Low fuel alarm
Additional outlets	Snow melt system
Additional lighting	Ductless splits
Pull down attic stairs	Water softeners
New water heater (tankless?)	Garbage disposal
Extended warranties	Insta-hot water heater
Service agreements	Towel heaters
Outside faucets	Radiant floor heat
Additional oil tank	Larger return duct
Gas fireplace	New ductwork
Bathroom remodel	Additional attic insulation
Pool heater	New deck? Etc., etc.

CHAPTER 12
The Patient

There are as many types of houses as there are combinations of equipment, maybe even more. As you travel across this vast country, you will see different styles of housing, different methods of construction, and probably some different ways to heat and cool a house that you didn't even know existed. For the purposes of this course, we again will stick to the basics. First, we'll look at some typical styles that you may come across in just about any town. Then we'll study some of the problem areas, or common mistakes, made with each of these. You may never see some of them, but they may help you overcome a challenge that I have never seen. Almost every single day that you work in this field, you will discover some new obstacle.

Let's start with the simple ranch. In my opinion, this is where all your bread and butter may well be made. The basic ranch, L-shaped, or raised, is one of the simplest of all houses. They can be built on a slab, or over a basement. The basement can be finished or not, and either way, it really doesn't matter, because it is mostly below grade and won't be a problem to cool anyway. Most importantly to us, they have a nice simple attic. Even if it's a modular, there probably aren't too many with really restrictive trusses. Trusses are the big wooden W or V in the attic that hold up the roof. You'll rarely find any real challenges in this house, as long as it hasn't been altered. The most common challenge in this style house is when there is a cathedral ceiling in the kitchen, dining room and living room. That's usually still pretty easy, unless you start to throw in a few skylights. Then the biggest challenge is throwing the air from the wall in the attic, all the way across the living room, so you don't get a warm spot at the end of the house. The first thing I would consider is downsizing the ductwork slightly, so as to increase the pressure and give you better throw from the diffusers. Other than that, put a return on that same wall, otherwise the hot air near the peak will never have a chance to escape.

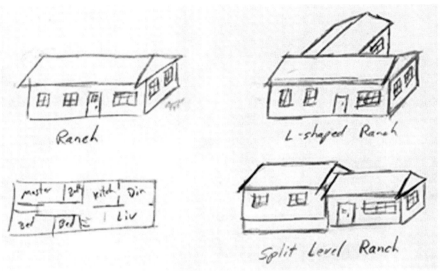

Fig 29

The next challenge, which again is a simple one, may come from a split-level ranch. Basically the same house, except that the bedrooms are three feet higher than the rest of the house. In this case, you'll usually want to place the air handler in the upper attic, because that is where you should put the return grill. What you need to watch for in this house is whether or not the two attics connect. In some, you can crawl from one to the other with no problem at all. Others don't connect, so what you will most likely have to do is drop your trunk line into the ceiling of the hall closet next to the stairs, then through the back wall of the closet, and you should be in the lower attic. Be neat about this, and the homeowner will be happy to give up the little bit of space, if it means a cool house. By the way, there may be no access to that lower attic, or there could be an access in that hall closet, above the stairs, or maybe in the hall closet by the front door. There's a good chance that nobody has ever been in there. Here's a great opportunity. Create a usable space for the homeowner with good access, and good light (charge for it, of course), and that should equal a good bonus for the add-on. Again, the ranch should be a good, profitable house for you, because, they are simple, there are a lot of them, and it should be easy for you to get good at.

As far as adding air conditioning onto an existing furnace and duc-twork here, the only problem you may encounter is on the split-level, and that would be the need to add a high return.

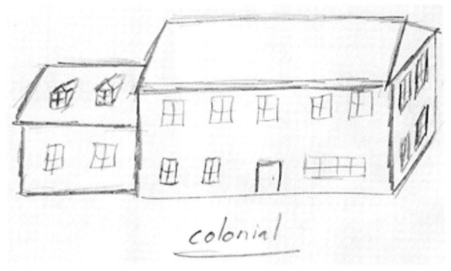

colonial

Fig 30

Our next victim, uh, I mean patient, would be the colonial (Fig 30). This is the next logical style, because it is very similar to the ranch. Almost as if there are two ranches, stacked one on top of the other. As you'll see, with each style of house comes its own group of questions. The colonial starts off easy. The attic is the same as the ranch, usually with more headroom. You may be able to stand up here. You will find three or four bedrooms, one or two bathrooms, and plenty of space in the hall at the top of the stairs. It's unlikely you'll encounter too many cathedral ceilings here, and even if you do, they won't be a problem. What may be a problem though, is the first floor. If you are doing this house with one system, getting ductwork to the first floor is the problem. The idea here is to use the closet in each bedroom for the duct to the room below. That works well, until you get to the master bedroom. Typically, you'll find one room below the master bedroom, that doesn't have a closet above it. If you can't run the duct down in a closet, then between the joists in the first floor ceiling, the only way to do it is to run the duct in the corner, exposed, and have someone box it in. The recipe to this job is to have, _IN WRITING_, who will frame and sheetrock, etc. Remember, if you are going to do it, that means at least three trips back with taping and spackling. Get paid for it.

Other than that, the only other problem you may run into would be the bonus room, which is the room above the garage. Don't discount this area, and make sure you include this room accurately accounted for in your load calculation. Chances are very good that this one room will call for more air than the whole rest of the second floor if it has a cathedral ceiling. And don't forget the return in there. Not putting a return in that room would be the same as placing walnuts on your head, and opening them with a baseball bat, only more painful. Try this at 10:00 on a Saturday night; "I paid $11,600.00 for this system, and it's 9 degrees hotter in this room! I want someone here NOW!!!" In all seriousness though, understand that this room heats up quicker than the rest of the second floor because there are more exposed walls and no attic above it. Make the homeowner understand that there may be a difference in temperature in that room if you can't get the proper ductwork there for whatever reason and *write that on the contract*. The best way to do a colonial is to do two systems. That way it will be more efficient, and the customer will have better control of their cooling. Zoning is also an option, and a good one at that. Zoning works very well in a colonial. You can use the same return in most cases, and you will have enough room in the attic. Oh, look, that bonus is another partial boat payment.

Okay, enough with the easy stuff, let's go to...The Cape! (Fig 31) I can tell who the experienced people are by the loud sighs. The danger here is that it looks like a simple job from the outside. Salesmen are famous for under pricing these jobs. Make no mistake about it, most capes are trouble, and for more than one reason. I know I keep stressing some of the same things, but if you want trouble free installs, listen closely. This house is all about return air. If you can't do with the return ducts what I am now about to tell you to do, either don't take the job, or write this sentence on your contract: 'This house will be, without any doubt, 10 degrees hotter upstairs, minimum, and you won't like it!" Then have BOTH homeowners sign it, and have it notarized! Then do the job, and hope for the best. If you are adding a coil and a condenser to existing ductwork, it will be worse than that. Can you see the battle scars yet?

If it seems like I am trying to scare you, you're right. These houses are that much of a problem. Let's break it down, and see why. Most capes that I have come across have little or no attic to work with. This is basically a ranch that has two added bedrooms in the attic. There are knee walls to work with that may or may not be feasible. If you can, put the air handler and ductwork in there, you'll be able to put enough supply and return air on the second floor, and may make it out of this alive. Downfalls here include; not being able to get to two rooms downstairs and a very tight fit for all of your equipment and ductwork. If you do have an attic to work with, then the problem becomes not having anyplace to run the ductwork down to the first floor, because all of the closets are in the wrong place. Unless, you can run the ductwork between the joists from the attic to the knee walls, which is almost always impossible, you won't be able to do it. And if you are trying to do an up flow system from the basement, think long and hard about this before you start. You **_MUST_** have the return duct high on the second floor. This is not an option. Notice the **_MUST_**! I know, there is no room to get it up there, and I also know that you can't even get enough supply air up there. If you can't agree on a place to run the ducts, maybe exposed somewhere in the dining room, walk away and don't do the job. I know what's going to happen here. Some of you are going to say that you can oversize the supply duct(s) a little (you can't), and it'll be fine without the return (it won't). Quit your job and go work at the post office.

This job will haunt you for years to come, and you will never get paid. Good luck, and don't say I didn't warn you.

Fig 31

Whew! Let's get away from capes for a while. The rest of the houses you may come across will all present their own little problems and obstacles, but the difference will be that you will probably expect them. When you go to a custom house with many levels, alcoves, cathedral ceilings, and other odd shapes, you won't be surprised at what you come across. You should also know that when you're in a very large house, think big. Need for capacity grows quickly here, and can be deceiving. The people who live in those houses know that their home doesn't call for a $5,000.00 attic system. Don't be afraid of the big numbers, those customers aren't. You'll be doing the right job for everyone involved.

Mobile homes, if you have them in your area, are a simple, quick, one-day job. They have a furnace in a closet, with a duct system that runs underneath the house. Most times with this type of house, you will install a package unit. That is, a complete system, condenser, and evaporator coil, all in one. You get a damper package to tie into the existing ductwork, put the unit in place, tie in the electric, and you're ready to go.

Victorians and other old houses, provide a number of challenges that you may or may not anticipate. Things such as plaster ceilings and walls, rooms that don't line up with the area below, few, if any closets, and lots of molding, among others. If you do this job, be sure not to underestimate the amount of time it will take to do this properly. If there are 10 holes to cut in the house, and you normally can do that in ½ day, plan on 2 full days for this house. Look at where the line set will run. It's not going to be a straight, easy run. It will probably take a day by itself. I'm sure you get the idea, work quickly, but be realistic.

Sunrooms are always interesting. I once had a room that was 14 x 16 with 10' ceilings, which called for almost 3 tons of air by itself. What usually happens there, is that nobody will believe you when you tell them that you need that much, so someone will "just add an 8"there". Until the customer calls back, that is. Trust the load calculation, do it right the first time. If you have to go back, you're giving away the profit. And besides, if the load calculation says to do it, or your supervisor says to put in two 12" ducts in that 10 x 12 room, do it! If it's wrong, it's not your fault. But if they say put in 2–12"s, and you say, "it's too much trouble, and he's wrong anyway", and you put in an 8", guess whose fault it is when it's wrong?

What about the roughed house? Many new homes are being built as 'air conditioning ready'. This means that the builder has put in some boots and diffusers, a return grill, line set, and maybe the electric. This works out pretty well most of the time, but it's important to do a load calculation, even though the ducts to the rooms are already in. Sometimes in roughed houses, I sized the unit to the existing ductwork from the builder, because that's what the customer wanted. What I did was to explain to the homeowner that we would size the unit to the existing ductwork, but that we wouldn't be responsible for how the builder sized it, and that it may not be balanced just right. After all, the builder doesn't care about his ductwork because he knows that any a/c problems will be blamed on you. A much better way to do it, I've learned, is to do the load calculation and confirm for sure where there will or will not be problems with the existing ductwork beforehand. If it's sized properly, great, no worries mate. If there is a problem area, you may choose to include it or charge for it, or just write it on the contract. The point is, do something with it. Don't ignore it. I tried it, it doesn't work. Build happy customers, they're more fun and they want to give you more money.

CHAPTER 13
Condensation or, Moisture Mayhem

Hopefully you know by now that one of the functions of an air conditioning system is dehumidification. This happens as the warm air from the house passes across the cold evaporator coil. The moisture in the air condenses, becomes water droplets, collects on the coil, and flows into the primary condensate pan. From there all we need to do is provide a leak free, properly pitched pipe to a drain line, or the outside, and let gravity do its thing. Sounds simple, right? Well let me tell you, I have had more problems with this simple concept then you could ever believe. If there is a way to keep those little, harmless drops of water from going down that pipe, I've probably seen it. Have you ever seen one of those cute little harmless drops of moisture take down an entire ceiling? Too many times, I tell you! When that little drop gets together with his friends, and he's got a lot of them, they become domestic, aquatic terrorists! And they do their best work at night. You would think that this would be as simple as letting water run down a hill, right? Nope. Those little guys go up; and across; and everywhere you think they can't go, until they are good and ready. Now they're going to go down, but not until they take a ceiling, you, and your profit with them! Except now you have this chapter, and I'm going to tell you how to foil their evil plans. (I really wanted to write a murder mystery, but this was the best I could come up with!).

By the way, the condensate problem is probably the most common complaint I ever got. Rarely, if ever, did I have a customer tell me that their house wasn't cooling properly when we didn't know there would be a problem beforehand. But the condensate problems kept coming up. Here's why.

First, how basic is this? You take off the cover to the air handler, and for whatever reason, the whole unit, or just the pan itself, is pitched the wrong way, I just can't accept someone telling me they didn't know that the pan had to be pitched toward the drain. Is it possible that someone can't figure this out? Okay, from now on, everyone hold up your right hand, and promise that when setting an air handler, you will always take your level, set it in the pan, and make sure that the bubble in the level is away from the drain hole. Everyone get it? Good! Now let's make sure that the little knockouts are removed from the drain holes. Not all of the units have them, but check anyway. Good. Now screw in the male adapter to the lowest hole available, and while we're at it, let's actually use glue to connect the rest of the PVC.

Yes, each one of these are simple, but very common, careless mistakes. Okay, now some companies put a clean out tee next, and then the trap. That's fine, but you *must* put a cap on that tee. I don't care what the other guy says, just listen to this. In some cases the unit you are installing may not be getting enough return air. So what you ask? And why not? We'll talk about the 'why not?' in a few minutes, for right now, let me tell you this. If the system isn't getting enough return air, it will suck air in through the top of the cleanout tee. Just enough to create a little vacuum, so as to keep the water in the pan from going down. I know you don't believe me, but it's my book. You write the book then, and I'll believe you.

If you suspect you are having this problem, try this; shine a light into the top of the cleanout tee and then shut the unit off. What happened? (This is just like science class). If the water released when you shut off the unit, you have a return air problem. Take a piece of clear plastic, and put it over the hole, and try it again. What happened now? If that's the extent of your return air, condensate problem, consider yourself lucky to have found it now. While we're on traps, we need to point out that if the trap wasn't there, we would have had the same problem. The water in the trap acts as a barrier to keep air from coming in and creating a vacuum. In other words, the blower can suck the air in enough to cause a problem, but is not strong enough to pull the water out of the trap.

This all has to do with negative pressure. Negative pressure is real, and if you don't acknowledge it, it's going to destroy your customer's ceiling, and their fondness for you. Here's the opposite situation. Did you ever see a system that was double trapped? One at the unit, then maybe one at a waste line? It won't drain. Cut the pipe by the lower trap, (bring a bucket), and watch the water flow. Damn physics.

If you thought that seemed a little strange, listen to this. I used to see a lot of media filters installed, and I didn't realize this for awhile, but some installers would mount the media filter to the back of the air handler (O.K., so far), then pan the filter off, cut a collar into it, and attach the flex to the collar. Anyone notice anything wrong with that? Here, look at the diagram (Fig 32).

What's wrong here is that by panning off the face of the filter, you just cut down on the airflow through the filter by about 35%. Now, already it's hard enough to get air through that thing, forget about covering half of it! Anyway, the result, again, is negative pressure. The fan in there may be trying to put out 1200 CFM, but you're only letting it have 700 CFM back. It's trying to find that other 500 CFM in the seams of the unit and the condensate drain holes. In fact, until that fan gets the air that it needs, no water is leaving that air handler. The way you fix this one is to add a box with a minimum depth of 12" to the end of the air filter. Now the air coming through the collar will have a chance to spread out and use the entire surface of the filter to pass through, rather than that reduced area before. Ta-da! No more condensate problems (I probably had to see that ten times before I figured that out.)

While I was searching for the cause of all these condensate problems, I came across something that had been in front of my face for years, but I just never saw it. You know those times when the salesman made a mistake, and sold the job too cheaply? And how, rather than putting in a 3½ ton air handler, you thought you could put in a 3 ton, set it on high, and still get over 1400 CFM out of it? Well, guess what? You can't always do that. Why? Let's go back to one of those charts in the manufacturers books (Fig 33, on the last page of this chapter).

Do you see how all the CFM boxes on high speed are gray? Do you know why they are gray? Look at the bottom of these charts. They usually say something like: **"Usage on the high fan setting <u>will</u> cause condensate problems"**. <u>**WILL**</u>. And this is not the Joe Blow air handler company. This is a chart from a major manufacturer!

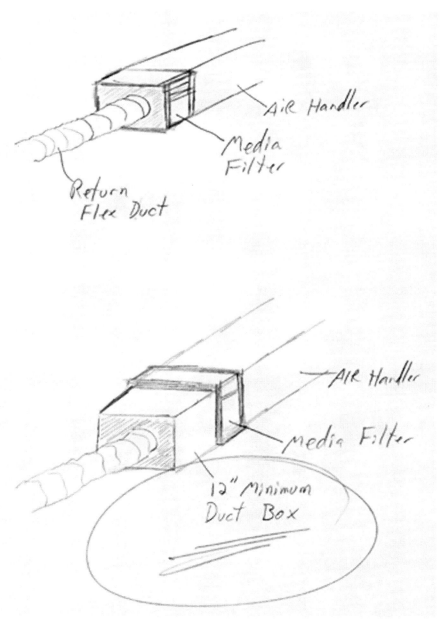

Air Handler

Media Filter

Return Flex Duct

Air Handler

media Filter

12" Minimum Duct Box

Fig 32

Why would they sell a unit that you can put on high, but can't use? It's one of life's great mysteries. And you know what? It doesn't really matter why, you just can't.

Other things to note with condensate systems would be things like; always put a secondary pan of the right size under the air handler, always run the secondary drain line out completely separate from the primary. That means no tees. The primary drain should go someplace inconspicuous like out with the line set, or into a leader pipe. You can put them into a vent or waste line, but some local codes prohibit that. Even worse, I've seen four or five vent lines that weren't connected by the builder, if you can believe that, so the condensate ended up draining into a wall! That's a mess. If you are going to drop the condensate line into an eave, be sure to have the pipe actually penetrate the eave, because if you don't, the water may follow the eave back into the house, and into the wall. Again, I've seen this too many times. Take the extra 2 seconds and do it right.

The secondary line should drip onto an area that is very obvious, like the picture window in front of the house, or the front door. You want people to see the water, if there ever is any. Remember, if water is coming out of the secondary drain, you know there is a problem with the primary drain and they need to call you. Another nice safety feature is the float shut off that attaches to the secondary pan. If that pan fills up, the float will shut the unit off altogether.

I once went to a house after a new installation that was, once again, having a problem draining. In fact, this was another one that overflowed the pan. I climbed up into the attic with the technician and went over everything we could think of. The air handler was properly pitched, the knock-outs were out of the pan, there was plenty of return air, but the water just wouldn't come out the end of the pipe. I had the tech blow it out with compressed air which proved that the line was clear. I was stumped. We kept searching, but found no answer. Then I told them to pull up all the attic insulation that covered the entire condensate line and then found the culprit. The PVC pipe was pitched about ¼ inch per foot for about 15 feet, which is good, but then there was a beam in the way. His solution was to add four elbows, creating an upside down

"U" over the beam, then continued with the properly pitched pipe. The top of the "U" was about an inch and a half higher than the primary drain pan. I asked the tech to explain why he had done this, to which he responded, "Well, I thought if the water went fast enough down the pipe it would just go over the beam". Needless to say, it didn't. Just to be clear, it won't; ever.

That covers the bulk of condensate problems. Don't take the lack of return air thing lightly; remember all this, and one day you'll be the hero.

Performance data

AIRFLOW PERFORMANCE (CFM)

MODEL AND SIZE	BLOWER MOTOR SPEED	EXTERNAL STATIC PRESSURE (IN. WC)											
		0.10		0.20		0.30		0.40		0.50		0.60	
		208V	230V	208V	230V	208V	230V	208V	230V	208V	230V	208V	230V
FA4A 018	High	660	725	615	675	565	625	500	565	405	470	---	---
	Low	585	650	540	605	490	555	420	485	345	395	---	---
FB4A 018	High	860	925	815	870	785	820	715	760	645	690	550	600
	Medium	650	740	625	705	585	660	545	620	480	555	385	450
	Low	565	650	535	620	500	590	460	545	405	480	330	385
FA4A 024	High	940	975	890	925	835	865	780	805	715	735	635	650
	Low	820	900	785	855	745	805	700	750	640	680	545	575
FB4A, FC4B 024	High	945	975	900	930	840	870	780	805	695	725	560	595
	Medium	835	900	795	855	745	800	690	740	610	650	470	510
	Low	605	695	575	665	530	625	485	580	425	510	340	395
FA4A 030	High	1075	1170	1030	1115	985	1055	920	990	850	910	750	805
	Low	825	960	810	935	790	890	750	845	690	780	590	680
FB4A, FC4B 030	High	1260	1305	1200	1245	1135	1170	1065	1110	985	1015	880	900
	Medium	1055	1170	1020	1115	980	1055	930	1000	960	920	755	810
	Low	830	950	805	925	780	890	740	850	685	790	595	700
FA4A 036	High	1320	1405	1265	1345	1205	1280	1135	1210	1060	1120	960	1025
	Low	1100	1215	1070	1170	1020	1115	960	1060	890	980	805	895
FB4A, FC4B 036	High	1485	1550	1425	1490	1365	1420	1300	1350	1230	1275	1150	1190
	Medium	1235	1380	1200	1325	1160	1265	1110	1210	1055	1140	985	1070
	Low	1035	1185	1010	1150	980	1115	940	1070	890	1010	825	935
FA4A, FB4A, FC4B 042	High	1580	1710	1540	1655	1495	1595	1440	1530	1375	1445	1290	1355
	Medium	1400	1570	1375	1525	1350	1480	1305	1425	1255	1360	1175	1280
	Low	1195	1375	1180	1350	1165	1325	1135	1285	1085	1240	1020	1160
FA4A, FB4A, FC4B 048	High	1880	1935	1785	1830	1700	1735	1615	1645	1520	1555	1430	1460
	Medium	1740	1840	1660	1750	1585	1660	1510	1575	1435	1485	1350	1390
	Low	1425	1605	1395	1555	1360	1495	1315	1430	1255	1360	1170	1270
FA4A, FB4A, FC4B 060	High	2145	2245	2085	2185	2030	2115	1965	2045	1905	1975	1830	1895
	Medium	2025	2175	1970	2110	1915	2050	1860	1980	1805	1905	1740	1830
	Low	1680	1895	1655	1855	1625	1810	1595	1765	1555	1705	1500	1645
FB4A, FC4B 070	High	2205	2285	2130	2205	2050	2120	1960	2025	1875	1930	1790	1825
	Medium	1880	2075	1845	2015	1795	1945	1745	1870	1675	1790	1595	1700
	Low	1570	1825	1560	1795	1545	1745	1520	1700	1480	1640	1420	1565
FC4B 033	High	1315	1385	1255	1315	1185	1240	1115	1165	1035	1080	950	995
	Medium	1045	1170	1010	1130	970	1080	925	1020	870	960	790	870
	Low	775	900	765	880	740	855	710	825	655	780	570	715
FC4B 038	High	1570	1700	1525	1645	1475	1580	1420	1515	1355	1440	1285	1360
	Medium	1215	1420	1180	1380	1150	1340	1110	1290	1060	1240	1000	1170
	Low	1020	1200	995	1185	960	1130	925	1090	880	1040	835	980
FC4B 054	High	1700	1835	1640	1760	1570	1685	1500	1605	1420	1520	1330	1430
	Medium	1505	1660	1455	1600	1395	1540	1330	1470	1260	1395	1175	1310
	Low	1300	1460	1260	1410	1205	1350	1145	1290	1080	1220	1000	1140

NOTES: 1. Airflow based upon dry coil at 230v with factory approved filter and electric heater (2 element heater, sizes 018 through 036; 3 element heater, sizes 042 through 060).
2. Not recommended for use above 0.60 in. external static pressure.

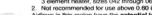 Airflows in this region have the **potential** for condensate to blow out of the drainpan. If usage in this region is desired, consult duct static pressure graph for allowable return and supply duct static pressures.

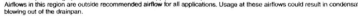 Airflows in this region are outside recommended airflow for all applications. Usage at these airflows could result in condensate blowing out of the drainpan.

Fig 33

CHAPTER 14
More Ductwork Stuff

As with most of the chapters in this book, this chapter could this be a book by itself. Ductwork really is a whole science and art form, in itself. Don't get me wrong, you don't need a PhD. But, anyone that has watched a career sheet metal guy, or even many of the duct board mechanics, knows that this is far from putting together a few boxes with insulation. As the main viaduct for a forced air system, there are many trials and tribulations to be experienced in retrofitting the residential house or building. This delivery system may reside in the basement, attic, crawlspace, or even through a finished area. While there are many components to the system we are discussing, the duct system will require, by far, the most planning, and design work. Not only does the duct need to deliver a specific quantity of air at a certain design pressure, it needs to distribute it to the correct area, keep it cool (or hot), eliminate or reduce noise, and keep it clean. All of this while fitting nicely into the bowels and black holes of the house that nobody wants to talk about. Yes, getting from here to there is often an art.

First we'll look at materials. For many years, galvanized sheet metal was the standard for all duct systems. Once in a while you'll see some aluminum. That's kinda cool. Sheet metal has many advantages in itself, but one thing in particular it lacks is an 'R' value. Actually, it does have one, but it's something like 'R-.0032'. I was thinking of something more along the lines of maybe 'R-8' or so, especially when you have A/C running through an attic, without some insulation you'll have an indoor rainstorm as I mentioned earlier. So you'll want to address that issue. You may have seen some white, bandage looking stuff on an older system. That's the famous asbestos insulation. You're not supposed to eat that, or make scarves out of it, just leave it alone.

These days you can order, or make your own ductwork with fiberglass insulation inside. Another option is to use duct wrap, which comes in rolls of different materials. This can also be used to insulate existing systems. Another popular option has become duct board. This is basically a board made of compressed fiberglass, usually treated with a tough acrylic polymer coating on the air stream surface, some with an antibacterial coating. The sheet metal guys frown upon duct board, usually because of some false info they've been given, or they assumed. The fact is if you can get a tin-knocker to give duct board a try for a month or two, they'll often buy into it. Reality says that there is a place, and call, for both. Sheet metal has structural integrity (it's stiff, and won't crush easily), duct board is light, and lends itself to easy transitions, and alterations on the jobsite. Metal is time tested, duct board is newer and high-tech. Metal has sharper edges; duct board has those little fibers you can never get off. We could go on all day. Insulated metal is usually required on commercial jobs. All in all my personal choice is…nah, I won't tell you.

In the beginning of this book, we talked about doing an accurate load calculation, and sizing the ductwork. Well, here we'll get into more of the actual construction. First, let's eliminate any discussion about the octopus. This is the system that has a "plenum", no trunk line, and all of the branch ducts spreading out from there. Yes, they may work okay in some situations, but you won't ever hear me say that. If you had a 3 year old son, and he wanted to go out and play in the snow, would you give him *your* coat, or would you get one that fit him? It's the same thing. You're a professional, do a professional job.

As far as I'm concerned, a load calculation's main purpose is to properly size the ductwork first, and the units second. This is where, if you do a proper duct job, you can build the balance right into the system. It's far better to have a long trunk line, than a million miles of flex. No, it's more important than that. Make it a rule to not allow your flex runs to be more than 12'. About now you should be able to get two runs out of each box of flex. With that in mind, if you have a

50' long ranch, you should be installing at least 25' of trunk line, not 10'. In some houses it will be necessary to do a 'T' configuration, or an 'L', or maybe some other letter of the alphabet. When making a 90° turn, you should use turning vanes, a sweep, or at least two 45Bs. A 90° will work, but you may get some surging, or excessive turbulence. You should strive for a passive system. That's the whole idea of variable speed air handlers. You don't hear them, you don't feel them, they just cool the house. You almost never know when they're on. Remember, that's the goal, cool the house without the customer knowing it. Incidentally, if you have a house with problem ductwork that you can't access to fix properly, a variable speed air handler may be able to camouflage some of the issues.

When designing the ductwork, refer back to the load calc and duct sizing sections for the details. Generally on a residential duct system, I find that for about every 400 or 600 CFM taken out of a trunk, you should transition to the next smaller size. For the average house, you shouldn't need any more than two transitions. Don't overcomplicate it.

For branch ducts, galvanized pipe has long been the standard, and still is in some areas, such as new construction. However, flexible duct is more common these days, especially since it's already insulated, it's quicker, and easier. It's useful in almost every job, as long as you follow the manufacturer's recommendations. These have to do with the correct collars to the trunk, minimum turning radius, and methods of support. Often times I'll see an attic system where the flex was strapped up with a piece of wire. Now the 8" flex is what, 3"? Let some common sense into the picture, it won't hurt. Flex duct is available with different sheathings, which will affect its life expectancy, so be sure to get the right kind for the application.

When cutting the branch lines into the trunk, a few rules here need to be addressed. Under almost no circumstances should any run ever be cut in at the end of the trunk. I mean the end where the cap is. There will be a few situations in your career where that may be a solution to a problem, but that's it. One or two out of maybe every five hundred. For that matter, nothing should ever come off the last 18" of the trunk.

There is turbulence in this area, and the trunk needs that 18" stabilize the pressure and velocity. Another area that you should stay away from is the plenum. Depending on the design of the ductwork, the air velocity at the plenum could actually create a negative, or backwards draft in a run cut in here. Avoid branch lines off the plenum if at all possible. If there is no other option, and you find that no air is being delivered to that line, replace the collar with a "scoop" collar. They work, but you'll be better off if you can move the duct farther down the trunk. Stick to the plan.

When it comes to areas other than the basement or the attic, you may need to look at options other than round flex or pipe. Oval pipe (galvanized steel) is available in 6", 7", and 8" equivalent sizes. Wall stack is another version of oval, but it is actually a rectangle. Both of these are meant to fit between the studs of a 2 x 4 wall. A wide variety of fittings are common, including 90's, flat 90's, oval to round adaptors, stack heads which are fittings to go from oval to a diffuser, usually 6 x 10 or, 6 x 12. Talk to your local supply house and they'll furnish you with a complete catalog of sizes and fittings that they carry. The same goes for metal duct, flex, collars, and every other piece you could possibly need.

It would be great if, in the retro-fit world of air-conditioning, they were all only attic or only basement systems, but reality states otherwise. Very often you are going to find yourself passing through one floor to get to another. You can do this with a variety of materials and cuss words. The trick has more to do with *where* you can run it. Closets are very common. If you stick to the front wall of the closet, you usually won't disturb the hangers, and you won't have to cut the shelf. This can be a hot topic with the woman of the house (closet space) so proceed with care. Before you do anything, spend some real time measuring and investigating to figure out where the holes in the floor and ceiling will end up and what obstructions may arise such as pipes and wires. Remember, for every wrong hole you drill you have to patch one. Discuss this whole process with the homeowner, if you can, so there are no surprises. I know the salesman or even you already did that, but this is often overwhelming for people and they may have forgotten or not understood the first time. Use a small bit to drill a pilot hole, then stick a piece of wire in it that you can easily find when you go back downstairs.

You will come across situations where you have one closet to work with on the second floor, but two back to back rooms to feed on the first floor (from an attic system). Many times there just isn't enough room between the floor and the ceiling to actually get any ductwork into, so here is something you can try. If the two rooms below each called for a 6" duct, run one 8" through the closet into the floor. Attach a collar right to the floor itself for a good seal. Measure out in each direction an equal distance, say 2 feet. You'll want to keep this as short as possible, but long enough to get into each room. Now cut the holes in the ceiling for the diffusers in each room. Make sure you're in the right bay! Now with sheet metal or duct board, seal the bay just past each hole, so that the space between the joists becomes your duct. If you can, caulk it with silicone. You'll end up with a 4 foot long "duct" with a diffuser at each end, being fed by the 8" duct in the middle. This is definitely not something to make a habit of, and most likely won't pass an inspection, but may get you out of a jam and the room should receive the air it needs without tearing up the entire ceiling.

Another version of the same concept is the satellite box. Now if this were a class on duct theory, this would not exist. There is only room for this in the real world when there are no other reasonable options. Or, we could say that this is not an approved method, but it works, and may solve some problems. There will be times that, for whatever reason, you just can't get a trunk line to that area of the house. Maybe it's a split-level, and the two attics just barely connect, or perhaps you need to cross a cathedral ceiling. Instead of committing a felony and running 50' of flex to each outlet, build a box, like the one shown below, and cut your branch lines into this. Then feed it with an appropriate size flex, which will become your extended trunk line. This is the only acceptable time to come off the end of the trunk. Add up the branches; maybe you need 300 CFM there. Go to the next size up, in this case a 10". You will lose a lot of air depending on how far you go. There will be a little trial and error to get it right. The main thing to remember is not to run those 6" and 7" lines there by themselves. You'll end up with no airflow at all. Absolutely zero. Don't try it, it's going to be difficult enough the way I said to do it. If you really get into trouble, you can look into duct boosters, but I'm not endorsing them.

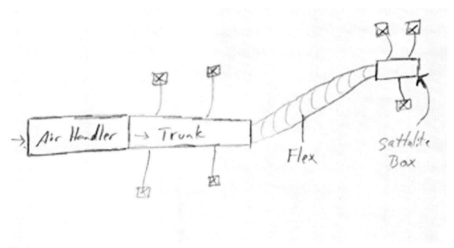

Fig 34

The return duct needs some lovin' here too, so let's talk about that. As I said earlier, one central return is common these days, but feel free to take it from as many areas as you think is necessary. In fact, when you have a room that is somehow separate from the rest of the house, like a bonus room, or an area where a door is always a closed, or even a room with a cathedral ceiling, do your best to get some return air out of these areas. Many installers make the mistake of assuming an area is going to be tough to cool, so they add more and more supply air, but overlook a return. That is backwards. You can only put so much air into a room. Pressure will build up, nothing will move, and it will still be hot. Believe me when I tell you, I have never been in a room that was too hot because there was a return duct there. Do you know anyone that has a pool? What if you pumped in clean, filtered water, but you never took any out? Yeah, you'd get some more clean water into it, but the gross, green, water would still be there. What's the difference?

Panning for return ducts was used for a long time, and still is in some areas. This is the practice of covering the space between two or three joists with a piece of sheet metal and using that space as a return duct. I prefer not to, if possible. The reason is that when I explain these systems to the customer regarding air quality, I say something like this: "What we're going to do, Mrs. Smith, is take this air right here, (as I grab some air in my fist), bring it up/down to the air handler, cool it, and put the same air back. We're not going to take any of the dirty air from outside, or from the basement/attic." If you do pan a bay, you are probably going to pick up a lot of that dirt regardless of how well you caulk it.

Here's an interesting one. You may come across a system that is noisy or just never worked properly due to lack of return air. So you look around and find a 30 x 20 grill in the wall. You pull out your trusty ductolator and determine that it's big enough for the system. But wait, there's no ductwork on the other side of the wall. Huh? Look at that, they used two bays between the studs as a return duct. It works in the basement when you pan the joists off, why wouldn't it work here? Well, the joists in the basement are probably 2" x 10"s. If the studs in the wall are 2" x 4"s, 16" on center, what size duct does that give you? A 3 ½" x 14 ½". Check your ductolator now, that's about 175 CFM at .05 WC (remember, it's a return). Oh wait; there are 2 bays, so that brings us up to 350 CFM. I don't care how big of a grill you put on the wall; 350 CFM will not replace the air you're trying to put out in any size system. It's amazing what you find out there.

If you have to run your ductwork through a finished area, you may consider using spiral duct. There are shops in your area that have a machine that can make continuous lengths of different diameter duct. You can get caps, tees, 90's, 45's and other fittings for the stuff. It looks good enough to be considered "finished" and may again be a solution out of a tough situation. Think out of the box. Whatever type of ductwork you choose, take care to properly seal all the joints and seams in unconditioned space. You can use mastic on metal, or tape on metal and duct board. If you do use tape, be sure you get the right stuff. UL listed tapes are available for each of the materials you work with, and *are* different. Numerous tapes exist for metal, which are different from ductboard tapes, which are different from flex tapes. Most are NOT interchangeable. Most come in a variety of widths, and if you buy them by the case, will even come with a squeegee. Yea! Ironically, standard, grey duct tape should not be used at all with ductwork. It will not hold up at all. Go figure. Again, talk to your suppliers to get the right stuff.

Much of the ductwork part of the job is dependent on your ability to solve problems. This aspect of the job is a direct reflection of your abilities. Take your time and do it right. Common sense is imperative. When in doubt, remember these few rules of thumb:

1) Trunk-long and straight
2) Flex-short and straight
3) Never take branches off the plenum, or end cap
4) Try to keep all branch lines about the same length
5) You can never have too much return air
6) Always brush your teeth before bed

CHAPTER 15
Electrical

When I started this book, my intention was to produce a guide that could be used by many to help improve some quality issues, and also to create a logical procedure where one didn't seem to exist. All the while thinking that most techs knew what had to be done, but just needed a little direction. However, what I found was that many of them really didn't know the correct way to do some of the things that we do. A lot of it was just a matter of practice, like cutting holes, or assembling ductwork. That's fine, but not when it's something that can potentially burn down the customer's house, or even kill you, such as line voltage wiring. There are so many things that should be discussed regarding electrical work, but that is a yet another book in itself. To begin with, where you work will determine if you are even allowed to touch the electric, therefore start with the local codes and your company policies. So let's hit the highlights and you can decide for yourself if you need additional training on the subject.

This subject is so vast, but we're going to start with the real basics. Wire. Very simply put, wire is how we get electricity from one place to another. The rate at which it flows, or strength of electrical flow, is measured in amperes, or amps. Conductors (wire) are rated by their ampacity, which is the amount of amps they can carry continuously under conditions of use without exceeding their temperature ratings. Voltage is the other factor in wire ratings. Most wire that you will come across will be rated for up to 600 volts, which is sufficient for the 120/240v that you will be working with. The amp ratings of a wire are critical if you want to avoid a fire, which I recommend. Fire is bad. Bad fire.

Wire size is a pretty simple concept, I think, but I see many, many problems in the field related to wire size. Installers with 10 years of experience are out there using wire way too small for what the job calls for. Wire is sized using the American Wire Gauge, or AWG. It is imperative that you match the wire size to the application. All manufacturers give you electrical specs with each piece of equipment, telling you how many amps that component draws, maximum breaker size, and often, what size wire to use. If you don't find minimum AWG sizes, use the following guide, which you need to commit to memory.

14 gauge wire - maximum 15 amps
12 gauge wire - maximum 20 amps
10 gauge wire - maximum 30 amps
8 gauge wire - maximum 40 amps
6 gauge wire - maximum 50 amps

You probably know that as wire gets bigger (more amps), the size number goes down. Yes, it's backwards. You can't go wrong by making the wire too big, within reason. (Don't run 6-gauge wire for your thermostat). You can get into a lot of trouble by making it too small. (Don't run 18 gauge wire to the condenser). Just as important as wire size, is the circuit-breaker that you tie to it. You must size from the breaker. In other words, if the condenser calls for 10-gauge wire and a maximum 30 amp breaker, you can't use 10-gauge wire, and then put in a *40* amp breaker, that would be wrong. The wire <u>must</u> be sized to the breaker. If you used the 30-amp breaker, and didn't have any 10 gauge, so you used 8 gauge, that's acceptable. Larger wire is okay, a larger breaker is dangerous. If the wire is too small for the breaker, it could overheat and burn up.

Along the same lines, you must look at wire length. Again, the manufacturer usually gives you not only the proper wire size, but also a safe length not to be exceeded. It may say use 10-gauge wire up to 75 feet. If it does not give you a guide, a good rule of thumb would be;

if the wire exceeds 100" in length, increase the size of the wire to the next size up. In other words, if you normally use 10-gauge wire on a 30-amp breaker, but this time the condenser is 140" away, increase the wire size to 8 gauge. Remember in wire world, bigger wire is a smaller number. But wait, does this mean we should increase the breaker size? NO! We still only want a maximum of 30 amps going to this unit; we only increased the wire size because we don't want the wire to burn up delivering the 30 amps. This is a very serious issue that needs proper attention on every job. If you need more training on this subject, tell your supervisor. Don't let this subject get pushed aside.

I want to stop in the middle here to say that all of this is subject to individual local building codes. While most of this is pretty standard stuff, your locale may have stricter rules that need to be adhered to.

There are many types of wire that you can use. The differences are mostly in the sheathing. Romex is the most common and easiest to work with. This is what is used indoors in most residential situations. It has a light plastic sheathing, which now comes in colors for easy identification. 14 gauge is yellow, 12 gauge is orange. Some municipalities require the use of BX or MC cable, which is metal sheath or armored cable. UF is another type of wire you should be familiar with. This grey wire is rated for outdoor or sunlight applications (UV resistant). They make UF for good reason, so don't run Romex exposed outside. It may look similar, but there is a big difference in its performance and longevity. Some local codes require all outdoor wire to be encased in EMT (Electrical Metal Tubing) or PVC conduit. As a matter of professionalism, conduit always looks better, and adds a feeling of quality to the job. Commercial jobs often require armored cable or conduit. PVC is easy to work with, as opposed to EMT which may require special tools and a little practice. Most codes require that when using any conduit, you only use THHN wire through it. THHN is unsheathed, individual colored wire, which *must* be used in a conduit.

When bringing any of these types of wire into a service panel, or junction box of any sort, you must use the appropriate connector. The purpose, aside from neatness, is to keep the wire secured so as not to pull out of the box, and also to keep the insulation from being cut and causing it to ground out. Make sure you use Romex connectors for Romex and UF, and BX connectors for BX or MC, and don't forget the plastic inserts on the metal shielded cables. Also take care to run the wire where it won't be disturbed, and secure it properly with Romex staples, clamps, wire ties or whatever else is proper for that application. This is an area that almost all techs can improve on. Look at your work when you're done, or ask a co-worker to look at it. Does it look like a 30-year electrical veteran did it, or like the local burger jockey did it? Better yet, would you pay for your work?

Running one continuous length of wire is always preferable, but if you do run short you can add a junction box, provided you secure it properly and put a cover on it. Junction boxes must be accessible, though. You cannot put one in an area that will be permanently covered by wall board or something similar. Let's talk about covers for a minute. Switch covers, receptacle covers, combination covers, blank covers, whatever. All of these covers hide, what looks to the homeowner, like an explosion in a spaghetti factory. Now, if you are trying to impress them with 'all that stuff', that's fine, but when you're done cover it up. This is another prime example of where guys take a perfectly decent job, and turn it into something sub-par, by carelessly forgetting the covers. Remember that the customer only knows what they see. If it works perfectly, and is exactly to code but looks like pig poop, it is pig poop. Covers cost, like, 2 cents apiece, so if you can't afford one, let me know, and I'll buy the next 20 for you.

Some companies use disconnects on the air handler and some use regular switches. They both are acceptable as long as they are installed properly. On the condenser, the service disconnect will probably be mounted on the house. That's okay, but don't just use a piece of UF from the house to the unit, use some seal tight with adaptors, or a pre-cut kit (Whip), or even conduit. Again, UF will work fine, but the Whip or conduit look so much more professional.

Each of the components in a split A/C system must be on their own separate, dedicated line. Almost all of these units are now 208/230V units, but there are definitely plenty of 110V units out there. More than one installer has gotten into trouble when changing out an air handler that was 110V with the new one being 220V, and used the existing wire and just changed the breaker to a double pole. Sounds good until you go to turn it on and realize that the bathroom, hall and master bedroom lights were all on the same circuit. Yuk. And while we're at the air handler, I think that adding a receptacle and a work light near the unit is a great upgrade or add-on, and should be sold on every job. In fact, many towns require it. So, go sell it and put it in, just don't pull it off of one leg of the 220V line (that's against code), get it from somewhere else.

Okay, let's run down to the basement. There are an awful lot of service panels out there. Many need to be upgraded. Some still use fuses. Did you ever get to a job, go to the panel, and find that nobody realized that this house has a two fuse, 30-amp panel, feeding the entire house? It happens. Other times the service may be acceptable, but the builder put every outlet on its own breaker, and there just isn't any more space for your equipment. Depending on the type of panel you are dealing with, you may have the option of condensing the breakers. These are half width breakers, sometimes called space savers, and are available for a limited number of brands, which can save you a lot of trouble and expense. Condensing breakers is an acceptable practice, combining circuits is NOT. When you condense breakers, you end up with the same number of circuits, but free up a few spaces. Naturally, each breaker will have the same amp rating as before. Don't confuse this with removing a breaker, and adding those wires to a nearby breaker. That is an absolute no-no. You may end up with all kinds of overload problems, not to mention violations. If the service panel is full, and space savers are not available for that brand, adding a sub-panel may be a viable alternative. Remember that adding a sub-panel does *not* increase the amount of amperes available; it only gives you more space. If the service panel has a 100-amp main breaker, and you add a 60-amp sub-panel, you still have only 100-amps.

Every once in a while, you will run across a panel that hasn't been made in 40 years, and parts or breakers are no longer made. In some cases the supply houses may have something laying around that may fit. Or, your customer's panel may not have enough amps to supply all of today's electrical needs like the new air conditioning system, so why not sell the upgrade? Most towns require an electrical license to do a service upgrade or panel change out, and your company may or may not have the necessary license. Don't let that keep you from getting the add-on. Enter the sub-contractor; you sell it, sub it, and collect it.

As you can see, there is an awful lot to comply within the electrical part of the installation. This is not an area to experiment on your own. If you are not qualified to work in this area, spend some time with someone who is and learn all you can. The more you know the more valuable you become. Yes, electric work can be dangerous, even deadly. But like anything else, follow the rules and you won't have any problems. Additionally, many opportunities exist for add-ons, such as light fixtures, exhaust fans, attic fans, receptacles (in the attic or other areas), sub-panels, and upgrades as we said before. The salesmen will miss most of these. You should be thanking them for all of the opportunities they have left you. Turn them into gold. Remember the boat payment?

CHAPTER 16
Finally, the Installation

They say that a proper installation is 90% preparation and 10% field work...no, that's not right. It's 99 and 44/100% pure...no, that's not it either. Maybe it's 2 all beef patties, and a partridge in a pear tree, and they all lived happily ever after. Well, in any case, the point is that there is a lot more preparation that goes into the right job than most people realize; getting the lead, setting the appointment, getting the sale and the financing. Then the survey, the load calculation, duct design, material list, gathering and loading the material, and getting a crew dispatched to the right house. Now you've met the homeowner, reviewed the contract with him or her, become familiar with the house, prepared the area with drop cloths, moved all the necessary obstructions, and made a plan. NOW, at last, you are prepared to begin. Ready, set, install!

That's a little anti-climatic. But it is the time to actually break out some tools and do the part that you want to do. It isn't my plan to tell you how to turn a screwdriver, or how to set up the ladder. The purpose here is to help you improve what you do now, point out a few things that you may not have realized, and, overall, help improve the structure and quality of your work. I am absolutely not going to get into what the flashing LED means, or how to wire a custom zone system. That stuff is better left to the manufacturers. Heck, we could be here every day for 6 months or more just learning the new stuff. Anyway, let's get to it. Let's assume that, for the purpose of this chapter you are installing a central air conditioning system in a basic ranch. Full duct system in the attic, media filter, 3 ½ ton air handler, and 3-ton condenser as the load calculation prescribed. This house is a little light on insulation in the attic. Sometimes, even if it is a basic ranch that you've done a hundred times, you get to the house, and set up, and you're just in a funk. Maybe it's Monday, maybe you just are not really sure where you want to start on this job. This is why it helps to have a regular system that you do over and over again. As we talked about earlier, you should have a staging area at the house to store all of the material, and a place to work rather than the living room. It's not a bad idea to go through the material list as you unload, especially if someone else put it together. If you're missing something, call and order it now so you don't have to wait for it when you need it. Next, the drop cloth and area clearing thing, etc., etc...

O.k., start with the air handler. Pull it out of the box, and flip the coil if necessary (you know you can do that, right?). Bring your drain pan up along with any 2 x 4's, 4 x 4's, blocks, plywood, and whatever else it is that you set the unit on in your area. You should already know where you want to set it, so build your platform or set up the hangers. Next, hoist the air handler up to the attic and set it in place. Don't forget the vibration pads/hangers. Before you do anything else, make sure this is where you want it, then go ahead and level the unit. Yes, that's it, the thing with the bubble. Good, okay. Now adjust it as necessary, don't just look at it.

One problem I often come across here is that the air handler is being set in the middle of the house, rather than toward one side when using a straight duct system. The rationale here is that the unit is close to the return, and close to the attic access for easy service. That sounds good, but let me throw this at you. When you put the unit in the middle of the house, the ductwork will cover less than half the attic. This means that half of the flex runs are going to be the proper length, but the other half will be much, much longer.

I actually came across this exact scenario recently. Not surprisingly, the rooms with the real long flex runs were noticeably warmer than the rest of the house. Now I don't mean 10° warmer, I mean 1 to 2° difference, which is usually within working tolerances, and deemed acceptable. However this customer was not happy. I spent some time there, and found that when this system came on, the rooms farthest away from the trunk with 30' flex lines were blowing warm air for about 20–30 seconds longer than the runs on the other side of the house, which of course makes sense. To complicate this problem, this system had somewhat of a short cycle (the system was oversized). So not only is there that little extra warm air in the beginning, and the fact that even when the system was 'warmed up', the supply air on those long runs was about 2° warmer than the rest, but now the system doesn't ever run long enough to compensate.

Now again, we're not building the space shuttle here, but those 2 extra degrees were in the master bedroom. The worst place to have a problem. If anything, make sure that room is colder than the rest. So anyway, three service guys have been out to this house, and now me. The customers have been very nice, but just want us to solve the slight difference. Already there are four or five hours on this call back and the problem is yet to be fixed. They all wanted to check the charge, or increase the size of the runs. That's where the profit goes! If the unit had been placed where it was supposed to, I am absolutely positive that there would have been no complaints. I can say this with confidence, because when I was in charge of six or seven crews or more, and doing many hundreds of systems a year, I would only hear, "this room is too hot", maybe 2 or 3 times a year. So move it down towards the end, <u>or</u> design a tee, then you'll have the best of both worlds. Yes, you want to put the equipment in a place that's easy to service, but it has to work first! Try to have the trunk line centered in the attic, not to one side.

Now that the air handler is in place, you should be out of that funk from before, and be ready to move right along. Bring the condenser to its new home with the pad, and move, cut, puree, or whatever it is you need to do to that bush, level the ground, and set the unit in place. Again, find the level (yes, the bubble thing), and try it one more time. Boy, nothing says poor quality from the street like a condenser on a 15° angle. Make sure that's not you. Now, with the two units in place, the customer feels like you are really moving along, and they will think you're great. You and I know you haven't done much yet, but they will think differently, and remember, it's all perception. Next, I like to run the line set up and across the attic. I do this for a few reasons. It may be easier to wrestle that copper across the attic before you have all the ductwork and other stuff in place, and also the weather may be a consideration. If you have nice weather now, why not get that out of the way? Who wants to be on a ladder in a rainstorm?

If your condenser is going on the end of the house, you can probably drill a 2" hole, (some techs drill a figure 8) stick the copper in, and make a nice long sweeping turn from vertical to horizontal. However, if you are putting the condenser in the back of the house, or for some other reason, you need to run the line set into an eave, or other tight area, try this; instead of fighting that stiff 7/8 line set on the ladder 30 feet in the air, hoping you don't kink it, measure the distance from the ground to the hole. Now add three feet, cut the copper, braze on an elbow on the ground (ACR, not plumbing, please), and attach the low voltage wire. Now, the line set will easily thread itself into that hole, without the threat of a kink, plus it will stay nice and tight to the side of the house. Nice and simple, with no flaring tempers. When tying in the line set, use long sweeps on each end if it will neaten it up at all. The extra 14 seconds will make a difference in how the job looks. Remember the details. By the way, if things are going well, this should all be done before lunch on the first day.

Now, I know that a lot of techs like to cut all of the holes first, and very often that makes the most sense. Consider this, though. If it is extremely hot in the attic you are in, you may want to do it this way, because you can get a vacuum on that puppy before lunch, and run some wire to get that bad boy running for *you* this afternoon. Why shouldn't you get some of that cold air?

Let's go back and do the ductwork next. Take a look at your duct layout. If you're still able to use R-4, or maybe even R-6, rather than having to look at it (the layout) twenty times, put a box of flex in each room throughout the house, matching up the prescribed sizes, of course. This way you don't have to think about it anymore. Now if you are using Airtec MV-3s or something similar, go ahead and grab a few, and put one by each box of flex. Pull the flex out of the box completely, and attach one end to the diffuser. Wait! Stop! Learn! More often than

I want to say, I find that some technicians, and most helpers, don't realize that while these items accept 6", 7", or 8" flex, if you are using the 7" or 8", you MUST pull the ring off the smaller collars. If you don't already know this, just grab the little tab on the side of the collar with a pair of channel locks, and pull off the little ring like a can of Spam. "Wow! That's cool!" Don't laugh; I've actually heard those words. Without pulling off the rings, you might as well just make every room a six inch flex. Call back city.

Okay, now go ahead and start cutting holes. When installing ceiling diffusers, they should be placed about 9"to 12"off of an interior wall (centered, naturally), not the middle of the room, and blowing toward the outside wall or windows. If those diffusers happen to be MV- s, you'll need to cut a 9" hole, and you should have two people doing this. Once you know where the hole will go, drill a pilot hole, and check it again. Make a 9" template. Then find a box that you can catch all the dirt and dust with in the room below, while the person with the template is in the attic cutting the hole. When you are done with that, feed the open end of the flex through the hole, with the tech in the attic pulling it up. Secure the four screws on the diffuser, and you just killed two or three birds with one stone. Boy, you guys are getting good! However, If you are using R-8, which you probably are, you won't be able to pull the flex up through the hole, so you'll need to drag all of the flex up to the attic. Then you'll have to crawl around and do your flexing from up there. If you follow my prescription though, you'll be working in a cool attic.

Now that you're done with that, you can split your crew up. Let's not have the two of you wiring the switch. I'm not telling you to be all hyped up like a tornado, but you shouldn't waste any time either, if you have to go to the supply house, just one of you should go. You're not on the buddy system here. Cut the helper loose, he can probably handle more than you think. Give him a shot.

Alright, so let's start dividing up the work. One man should work on the trunk line. Do it like the plan says. If you need to make a change, call whoever did the layout. I know you don't think you need to, but if there is a problem, and you did call, you will have nothing to worry about. Anyway, put it together, mount it, hang it, transition it, burp it, whatever you need to do, make a decision and do it. A lot of guys get stuck on something and they just stop. I mean they are done for the day because they don't know what to do. No matter what problem you have, either get the answer NOW, or move on to something else. Do something! While that's going on, the second man can mount the disconnect and whip and start running the line voltage wire back to the service panel. If you need to run any exposed wire outside, use conduit. PVC is cheap as dirt, and very easy to work with. The quality difference between conduit and UF is night and day. It's the little details, again, that will justify your high prices. These details are a funny thing. The customer may not notice any of them, but if you don't do them, they'll stand out like a sore thumb. Count on it.

Back to man one. He can now cut the flex, and tie it into the trunk in the appropriate places. Make sure that all the nylon straps are all tight, and you should also tape them. When working with flex, in a perfect world, all the runs would be the same length, without any sharp turns, etc. that's not practical in real life, but if you can get close, that will be better than government work. Next you may want to run the condensate lines before you do the return, due to lack of space. Condensate lines, I'm starting to feel, are the most critical part of this operation. It's best if you can run the primary out with the line set. If not, look for all other options before going into the vent, because that's often against code. You'd be surprised to find out how many vents aren't actually connected to anything, either. Even if they are, I have heard complaints about the dripping sounds when piped into a vent line. Next, run the secondary drain line completely separate from the primary. Absolutely do not tee them together. What's the point of even having a secondary drain if you're going to do that? The secondary should drain someplace obvious, like onto a window or door, because at least in theory, it should never see any water. Why, you ask? Because if there is water in the secondary drain pan, there is a problem, and the customer won't find out about it until it's probably too late. That's why we want it to drain someplace unusual. (Yes, on to an outside window or door! Did I have to say that, really?) You should also consider using drain pan float switches. Take every precaution you know of to prevent any type of condensate problem. See the section on condensate drains again. When laying out the PVC in the attic, you should use purple primer. Not just because the glue will stick better, but also because, in a quick glance, you can tell if you missed any joints.

By now the wire should be making its way up to the attic, if you haven't done it already. Depending on the house, whether the basement is finished or not, and a variety of other factors, getting wire to the air handler may be simple or not. In some houses there may be a chase to run the wire up. You may have to run it up with the line set (make sure it is UF), or you may have to run it up in conduit. Air handlers come in both 110V and 220V. Regardless of which, be sure that it has its own dedicated circuit (more on this in the chapter on Electric). Before you do the return, run the thermostat wire and this is as good a time as any to wire the switch and air handler.

If you haven't cut the hole for the return duct yet, do it now, and mount your grill. In some houses the access to the attic is so small that you will need to cut the return hole before anything else just to get the air handler, or even you up there. Now we said this was a 3-½ ton air handler, so don't use a 14 x 30 grill because you may get some noise, or a hum out of it. As a rule, anything over a 3 ton should get something in the neighborhood of a 20 x 30. A lot of guys don't like this because you'll most likely have to cut a joist into the ceiling, and frame in for the duct. If you aren't good at toe nailing, get some joist hangers, the kind you used when you built that deck. Very simple. Never pan the joists above the grill and stick a collar on that. Not only does it look like the dog did it, it may cause other problems, like noise. Always put at least a 12" box above the return grill. Another little item that you need to get into the habit of doing is to keep a can of white and a can of black spray paint in the truck. Look up into the return grill; can you see anything other than black? If you can, paint it. Carefully! Now build another box onto the return side of the air handler, again minimum of 12", and tie in the side return flex.

If you are doing a colonial, or some other type of two-floor house, and a system is being installed for each floor, most of the second floor installations will fit like a glove in the attic as we just went through. And although there are definitely some substantial differences, which we'll address in a minute, most of the basics also apply to the basement system. Let's take a look at what those differences might be.

Working in basically the same order as the attic system, the first thing on the agenda will be placing the air handler. Suddenly, a bunch of new considerations come into play that you didn't have to deal with upstairs. For one thing, space is usually of primary concern in the basement. In some houses, the basement is only used for storage and the like. In others, it's just another room in the house. With the first, you have all the clutter and years of junk to climb over, and work around. In the other, there usually isn't so much stuff, but everything is finished, and you have the problem of trying to run the ductwork. Then, of course, you have the planners. The planner is "gonna" move that wall, carpet this area, put a bar over there, and turn this room into a 12 lane bowling alley, complete with pro shop and arcade. So now you have to climb over all the stuff, but still can't put anything anywhere, because they're "gonna". Well, we're "gonna" put the air handler somewhere, and they'll just have to move the Olympic size swimming pool with the strobe lights and disco ball.

In following the same configuration as in the attic, we can leave the air handler horizontal, and hang it from the floor joists. The simplest way to do this, is to get a hold of some unistrut (also called Kindorf), cut two pieces about 24" wide, or 6" wider than your unit, hang four 36" lengths of threaded rod, slip on the unistrut, add washers and nuts, and you have a place to hang your air handler. If you are going to use a condensate pump, you may want to make another platform below the air handler using a piece of plywood and the extra threaded rod hanging on one end of the unit. Cut off however much of the rod you have left. If you don't you'll soon know why you should have, evidenced by the holes in the top of your noggin.

The other option you have here is to stand the unit up, as if it were a furnace. Now, although there are some exceptions, you usually can't cut the return air into the side of the air handler, like you can with a furnace, so you'll have to have a stand made for your air handler. Have a sheet metal box made, the same length and width as the air handler, capped on the bottom, about 20" high and with a ¾" lip on the top. Remove the cap from the bottom, screw the box to the return side of the air handler, and replace the cap. The thing you have to watch for here is the overall height, remember you have to put a plenum on the top of this, and you are quickly running out of headroom. You should leave at least 12" for the plenum.

In many situations, a 'T' configuration works well in a basement, because your ductwork is only about half the width of a straight system, placing the air handler in the middle of the basement. We'll get back to that in a minute, but for now I want to point out that unlike the attic system, the return duct may be your biggest obstacle in the basement, so make sure you consider that when you are planning the placement of the air handler. The two problems that usually come up with the return are; 1) how do we cross 'the beam' and/or stairs, and, 2) there is a window in the way.

Back to the supply side, when planning your ductwork here, the idea is to keep adequate headroom. Eight by something duct usually works best here and I'll tell you why. The center beam that runs the length of the house is usually about 9" high, and that is the best place to run the trunk. They were planning to box in the beam anyway, when they put in the 85 seat movie theater, with surround sound, now all they have to do is make the box a little wider.

With the trunk line run next to the center beam, you can run all of your flex between the joists and even over the top of the beam which works out great, just make sure that you run the trunk long enough so that you don't have to cross any joists with the flex. Now that you feel confident with basement systems, let's throw the modular house in here. "What's different about the modular house?" you ask. You say that you know most modulars have trussed attics, but what does that have to do in the basement? Hmmm. Do you know what one of the selling points of modular homes, and new styles of construction, is? No center beam in the basement. They were going to have a nice clean ceiling in that room, but now they may have to lose that, so make sure the customer understands exactly what they are going to end up with. If you show them what it is that you need to install, often times they will help you lay it out, and by involving them in this process, you are less likely to get a complaint later. Many of these types of complaints come from the customer not having any idea of what you have to do down there. You'd be surprised how many people think that you install the unit, and magically the air moves around the house without any kind of duct system.

Alright, let's move onto the diffusers. The outlets on a basement system are usually placed on the outside walls below the windows, unlike the ceiling diffuser upstairs. Basically, you cut the hole in the floor, install a boot which gets the flex or round duct attached to one side, then the diffuser drops into the top.

Cutting these holes requires more thought than the holes upstairs in the ceiling, because holes in a hardwood floor or ceramic tile are not so easily repaired, as is sheetrock. Take your time and do it right. You'll want to keep the diffusers out of the traffic pattern as much as possible, too. When it comes to the return grill, you must involve the homeowner. Although the grill itself can be smaller than the return grill you would put on the same size attic system (see specs from J&J, Hart & Cooley, or Lima), it is still going to be a very substantial alteration of their floor. I like to look for a central area of the house with a piece of furniture that may camouflage it. Make sure they understand that they can't put a carpet or a couch over it, but to put an end table over it would be okay. Explain the purpose of it and they will be more accommodating.

As far as what to cut the floor with, whether you use a sawzall, a circular saw, or something else, I'll leave that up to you. I will say that I have found when cutting tile floors, the best tool to use is a dry diamond blade for a 4"grinder. However, this will create a serious dust storm, so take time to contain it before you start cutting. No matter what you use, you will get plenty of dust, but at least this will get through it pretty quick.

Floor grilles are most often seen in brown. There are a great number of alternatives out there, but don't be mistaken about the price. A good solid brass return grill can cost you $400.00 or more. Of course you can get them in cast iron, aluminum, wood, or anything else if you are willing to pay enough. Most customers aren't that concerned with supply diffusers, because they are small and usually inconspicuous, but every once in a while you get the customer "who will die if you put those ugly things in my house!" Bonus time! Upgrade, upgrade, upgrade!

So, now you have the unit placed downstairs, ductwork finished, holes cut, condensate pump and tubing connected. From here, set the condenser, run the line set and wiring. By the way, when running a line set between the joists in the basement, be sure to secure it well, because they can occasionally have a slight rattle or buzz to them. Check the insulation again too, because you don't want condensate dripping on to the ceiling if they ever really do get around to finishing the basement.

If I'm not mistaken, all that should really be left is the leader pipe to cover the line set (if that's what you use), mount the thermostat, turn on the breakers, evacuate, open the service valves, and prime the trap. Keep in mind that If a typical attic system takes you two days to complete, you will spend at least three days on a basement job. So, it looks good so far! Cross your fingers, and start it up. You are now in the homestretch!

CHAPTER 17
The Homestretch

It's been two or three long hot days in the attic, and even more in the basement, but now you're in the homestretch. The finish line is in sight, and it's almost time to put another hash mark on your truck, or however you note how many systems you've completed. Everything is wired, mounted, and almost ready to go. Let's not miss anything, or rush the last few hours. You've done a good, quality job up to this point; just stay on track and you'll be successful.

By now, you should have a vacuum on the system. This is important to the life of the system, so don't shortchange your customer and leave it on for only 10 minutes. Learn the proper procedure using nitrogen, etc. and do it. Since evacuating correctly will take a while, don't wait until the last minute to do it. When you are finished, go ahead and open the service valves, and don't forget to put the caps back on. Now fill the condensate trap with water, and start the system.

Once everything is running, take a look at everything you have done. Check the charge now. Is the ductwork all sealed properly? Are all the nylon straps tight and cut? What about the condensate? Is the cap on the tee? Go check the end of the condensate line to make sure the water is actually leaving the building. It may take a while for the condensate to actually reach the end of the pipe. Remember about the condensate leaks being one of the biggest complaints? Spend some time here, and make absolutely sure that it works properly. If the inside of the plenum is getting wet, there is a problem. Figure it out, and don't leave it that way. There is a good chance that the negative pressure problem has come up, and is not releasing the condensate. I know that I have mentioned this a number of times, but this is a serious if not common, problem, and is definitely not limited to the inexperienced.

When you are confident that you have conquered the whole condensate thing, move on and check the rest of the system performance, including switches, thermostat, return, airflow, and the like. If balancing is part of your game, go ahead and do that now. If not, go to each diffuser, make sure the dampers are open, and that everything seems to be in order. If you do need to damper down a particular duct for some reason, it's best to do it at the trunk line, rather than at the diffuser. Doing it at the diffuser will cause unnecessary turbulence and noise, which you don't want. It may not seem like much noise now, but at 1:00 in the morning, when the house is completely silent, that little noise may be sound like a freight train.

Next, head down to the basement to check the electrical panel, make sure you turned everything back on, and replace the panel cover, and anything else that you may have had to move. Now, program the thermostat, and check again that it works in all modes (cooling, heating, fan, etc.). I shouldn't need to say these things, but many callbacks start with "now my heat doesn't work!" You know that it's something stupid that will take you all of 12 seconds to fix, but now you've already left and the customer is annoyed. So check the little stuff anyway.

Once all is in order, you can start collecting your tools. It never fails, if you put your tools away before you check everything, the tool god whacks you in the head and makes you take something back out. It's a law that can't be broken. Now pull out the paper towels, and the Fantastic, and clean up the fingerprints. They are there, so you can't skip this step. There is no way that you got through that whole job without a smudge. It's okay, just clean them up, and if by some miracle you can't find any of yours, clean up some from the kids. Your customers will love you for it. Then return any furniture that you may have moved, and don't forget the painting from Aunt Mavis that goes in the hall. What was that a painting *of*, anyway? All right, pick up the drop cloths without dumping all the dirt onto the floor, and sweep or vacuum, if necessary. I think it's a good idea to do anyway. These are the little things that are going to put you on the customer's side, and if they have any problems, they'll be reluctant to call because you were so nice, and they won't want to get you in trouble.

Next, take a walk with the customer around the house like you did before you started. Show them everything that you did, how everything works, and whatever it is that they have to know. Go over the filter with them, where it is, and what they have to do with it. Show them the condensate drains, and tell them what is supposed to happen and what isn't, like the purpose of the secondary drain. Now, go to the thermostat and explain how that works. These are all potential problems that can be avoided just by explaining them to the customer first. You should create a checklist of quality points, and things to cover with the customer. By doing this, you won't forget anything important. Then you should sign it, and have the customer sign it. Give a copy to them, and a copy to your office. Everyone will appreciate that. Don't forget the owner's manual and warranty paperwork. Add a little value by putting all of that stuff in a folder with your company's name on it. Now they really feel like they are important to you. Collect the check, give them a sincere thank you, and you are done! Congratulations on a successful installation!

CHAPTER 18
Quality

Have you ever been to Disneyworld? I am lucky enough to have been there a couple of times. Why is it that Disneyworld is one of the most popular vacation spots? Because it's an inexpensive vacation? I don't think that's the answer most people would give. So then, why is it that everyone loves to be at Disneyworld, and seems willing to pay almost anything to go? Not long ago I took my kids to one of the competitors, Six Flags to be exact. We had a great time. It was clean, fun, diverse, and big to say the least. You know what? It wasn't Mickey's place. Not even close. Why? I'm not exactly sure; it's just a feeling. It's like buying a one-year-old car from a dealer. It's spotless, no signs of wear, runs perfectly, no different from this year's model. Now get in a brand new car. It's different, it's better. I don't know why, it just is. Quality isn't just a nice solid, shiny, new unit; it's all the little things you don't notice. It's all the little complaints you *don't* hear. It's getting through the entire amusement park, and realizing that you never, ever saw any litter. When I say Disney was clean, I mean not a piece of chewing gum anywhere! It's as if the cleaning people are disguised as tourists so as not to be noticed. The people that work for Disney have figured out how to make this a truly magical experience. You never see any chipped paint; cigarette butts on the ground, overflowing garbage cans, none of that. Quality is something so good, you can't even think of what could be wrong, because there is nothing to remind you. Quality can mesmerize you without even knowing it. Let's learn to mesmerize our customers.

In our world, quality starts as soon as you pull up to your customer's house. What is their first impression of you and your operation? Is your truck clean and quiet? Are you clean and neat in both your hygiene and uniform? When you introduce yourself, do you look the person in the eye with your head up, a smile on your face, and a loud, clear voice? That's when quality is going to start for you, the installer. The chapter on Set Up, among others, touched upon quality. While every employee should be concerned with this automatically, most need to be trained to live it. We still need to keep ourselves 'in check', if you will. That may mean that your company has a quality control officer that keeps watch over the details, or more likely, you are left to police yourselves. Keeping tabs on your own level of quality can work, if you have a procedure, but it is still very difficult. As much as you like to think that you have it under control, I will bet my shoes that you can be a lot better. One reason for this is that once you do something like run a wire across the attic for the 7,123rd time, you are probably thinking about what to do with the neighbor's dog that keeps eating your tomatoes, not the thermostat wire in the attic, draped across the beams. Remember, quality is *not* being able to find anything to complain about.

Half the time the stuff you're working on is such second nature; you may have just wired the air handler and not even realized it. That's why it's very difficult to scrutinize your own handiwork. No matter how good you are, you can't ever see the job like your customer does. A good place to start maybe a checklist of problem areas that you and/or your manager may have identified. Now don't get defensive here, this really does apply to everyone. You, me, the owner of your company, all of us. We all need something to remind us to stop and look at what we're doing. Don't believe me? Okay, how many drop cloths do you have on your truck? You can't cover all of what you need to with only two. Booties? Where are they? Did you put a level on the condenser? Did you weigh in the gas? Did you watch the condensate come out of the end of the drain line? How is the Romex in the attic secured? You say it's only the attic. When the homeowner puts his suitcases away, and they see the wires just hanging there, he's going to think you did a sloppy job everywhere. What about the ends of the zip ties? Did you pick them up, or throw them in the corner? Want to make a $10,000.00 job, with a variable speed air handler, 18 SEER condenser, media filter and custom diffusers look like Pig Pen and the Mad Hatter did it? Don't secure the wires! That'll do it.

Look, you at least need a checklist to guide and remind you. If you use one, and give a copy to your customer and manager when you're finished, your quality rating will be so high that customers will begin to request you to do their installation. It will happen if you pay attention. If customers are asking for you, and your call back rate is dropping at the same time, you'd better believe you can ask for more money and get it. In addition to that, the quality of the entire company will increase, because the other installers will see how much praise and success you're having, and they'll try harder too.

I mentioned the call back ratio of your jobs. Get a list of all the callbacks on your jobs for the last six months. You probably aren't even aware of most of them. Many companies are split into departments such as; service, installs, etc., and if that's the case where you are, the service techs are most likely handling the callbacks. So what are these callbacks? I'll tell you without looking. Ready, my eyes are closed... most of your call backs are from sloppy work, because you didn't go back and really look at what you did, or you didn't test it. That's it! Get the numbers. I'll bet that almost every single callback on your jobs was from not checking your work. Go look, I dare you.

Once again, there is no big secret between what you do now, and what it takes to make you one of the best installation technicians in the industry. I am completely serious. Install what's prescribed, be neat and courteous, and make sure it works when you leave. It seems pretty simple to me. Most, yes most, installers seem to miss out on all of these. First, they think they know better than the load calculation, and make their own 'adjustments'. Then they are only fair on the neatness thing, they leave fingerprints by the thermostat and attic hatch, leave some garbage in the attic, and last, they don't look at the finished product. I just don't understand the last one, but it is so prevalent in this industry. I used to tell my guys, "Listen, when you're finished, I want you to hang out here for an hour or so, enjoy the air conditioning, and just make sure it all works." I didn't say vacuum the attic, or rake the lawn, or wash the customers car, I said hang out in the air conditioning. They wouldn't do it. That's why that top level of quality isn't achieved. Imagine that you may be able to bring your department from break even, to profitability, by just hanging out and watching things work? And, they're paying you to do it! Sounds like a plan I could follow.

Try this trick; sign your name to your work. After all, this is your work of art, be proud of it. That's right, grab your Sharpie and sign the air handler. Are you ashamed to, or are you proud of what you see?

CHAPTER 19
Customer Complaints

To not ever have any callbacks, is not at all realistic. But with proper management and training, you can learn to eliminate a good deal of them. I think you'll be surprised! Start by getting a breakdown of all your callbacks over the last six months. Look it over, do you see any patterns? You'll probably see the same problems over and over. That's good! Now, you can fix them. In spite of your best efforts though, some customers will still complain no matter how good you become. Don't get discouraged. If you take care of the next five problems areas, I'll bet you cut your call backs and customer complaints at least in half, maybe more. Again, this is where the profit goes.

1) Explain design parameters:
One of the first problems I see is the customer's expectation of performance. If the homeowner never lived with central air conditioning, as is the case with most retrofits, they don't really know what to expect. It is best if you or the salesman handle this before you start the job. The customer needs to be told that the system you are installing is meant to cool the house to about 20° cooler than the outside temperature, or whatever it is that you design for. If you don't explain this, your customer will call and complain that, "last week when it was 119° outside, I could only get the house down to 85° , and I want it to be 48°. Don't laugh; these are the kind of calls your office gets. Your customers need to have realistic expectations for performance. Put it in writing, it will help.

2) Explain how to work it (whatever *it* is)

"Nobody showed me how to work that damn, digital, programmable, nuclear thing on the wall, in fact, I think it bit me!" Make believe all of your customers are Idaho potatoes. That's right, you can't blame them, they weren't made to think, they're supposed to be in a salad. You must explain how to use the thermostat, where the filter goes, what the big thing outside is, and why you shouldn't crash into it too much with the lawnmower. They need simple, direct instructions. And by the way, <u>they</u> get the owner's manuals; you have enough in your truck.

3) Improper ductwork

A common complaint I hear is that they aren't getting enough air in 'that' room, or one room is too hot. Now, I take duct sizing and layout pretty seriously, so If I designed it, I don't usually believe it until I see it. One of two things always happens. 1) It turns out that room is exactly the same temperature as the rest of the house, they just "couldn't feel the air blowing". This is fairly common, so you may want to *show* the customer that all the rooms are the same temperature before you leave. Or, 2) It truly isn't working the way it should so I go up into the attic to find that the ductwork was nothing like I designed, and the run to that room is a 35', 6" flex, 6 turns, and tied to a beam with wire. Where do you want it, in the stomach or the back of the head?

4) Condensate leaks

This one is serious; water. Unfortunately, condensate leaks are way too common. What's worse is that they cause a lot of damage, and make for a very unhappy customer. And again, there goes the profit. Watch the problem areas I pointed out in the condensate chapter, and I think you'll have it covered. Make it a point to stick around for a while, and make doubly sure everything is working properly. Remember, the big drain pan under the air handler should never have water in it.

5) Someone didn't clean up

This last complaint is based on carelessness, or just plain foolishness of someone on the install crew. Not cleaning up properly. Not only does it look unprofessional to the customer, if they call the office, they aren't calling to say something is wrong with the system; they're calling to say they have a problem with YOU! And you can be sure that the boss is going to remember that at raise time. One thing to realize here though, is that the homeowner may have a much higher standard of cleanliness and organization than you do. You may have actually done a very nice job in cleaning up; nevertheless, they could be a little fanatical. I don't have to tell you they're out there. But, they're also laying out a lot of hard earned money for your work, so play the game, and if there is a question about it, talk to your customer before you get paid and leave. Better now than later.

Ask Yourself These Questions

1) Is this issue covered on the contract?
2) Did you do your part?
3) Did the customer do their part?

CHAPTER 20
Permits

If you work in multiple towns, cities, or villages, you may be required to hold numerous licenses, and/or take out any number of different permits, depending on the job you are doing. Often, only the owner of the business needs a license, but more and more towns or even states are requiring the techs themselves to hold them. As I said earlier, you are far better off playing the game, then you are questioning why. This is especially true when an inspector visits a jobsite. To me, why is a very reasonable question when dealing with local officials, because so little of what they do, or require, ever seems to make sense when it comes to inspections. As much as you want to explain to them why they are wrong, or why "that" shouldn't apply in this case, hold your tongue. You have to look at the complete entity at hand here and not just the official.

Do you realize what is involved in order for this municipality to request this $20.00 permit? A lot more than the guy with the badge standing in front of you. He's just the person that makes sure you do what the book says, of course it's usually his interpretation. If you're lucky, he'll have some common sense, and a decent knowledge of the subject. If not, just give him what he wants, and keep your mouth shut, even if it doesn't make any sense. Let's keep the goal in mind. Finish the job in a reasonable amount of time, satisfy the customer, and get paid. If you piss off the guy with the rubber stamp, what happens? You don't finish on time, then the customer is angry, and to follow suit, you don't get paid. Not to mention that the next time you go to get a permit in that town, or do any work there, they will scrutinize and question everything you do. So again, play the game, keep quiet and say "thank you". If they tell you that horizontal ductwork has to be installed on a Tuesday, and painted blue, adjust your schedule, and ask them, "Sky or Navy blue?" I'm not at all kidding.

The big guy, or successful company, is often the target of envy or suspicion, or both. If you are from out of town, or different than your competitors, be prepared to do everything better than by the book, even when the others aren't asked to. You may be asked to take out a plumbing permit for your condensate drain. Again, just do it. Some ways to combat this, though, are to start by having the same person go to town hall to take out the permits each time. By using the same person, it doesn't matter who, they will start to build relationships with these people, rather than having just the big company image with different people every day. Some small town people are put off by the big corporations. Remember perception is reality. Justified or not, they make the rules.

Another way to help your cause is to make it easy for the people who work there. Most of the time, it seems, you know more about the permit process than they do. It's been my experience that every time I went to a town hall, the procedures changed. Not because it really did, it's just that there was no actual procedure. Well, write up the procedure for them. Do it humbly, with their help. The trick is to not let them feel as useless as it may seem they are.

I've had to deal with an untold number of municipalities and permits, that seemed to grow larger in number every day, and sometimes, I think just because they knew 'we' were working in their town. Anyway, what I did was to create a master permit book in which I had a copy of every form, and every requirement for each town. I also made up a procedure sheet which would say what we needed a permit for, how much it cost, what forms were needed, how many copies, who needed to sign them, and anything else that may apply. Eventually, it all moved onto a computer. If you can get this procedure in place, I know it will save you some time, and possibly even keep you from an assault charge.

CHAPTER 21
Where Does This Career Lead?

Ever wonder where all the old installers go? Do they all retire as billionaires to the Caribbean? Maybe. What about you? You are most likely in your twenties or thirties, and you may not have given this subject a lot of thought yet. Trust me, you should. As rewarding and interesting as the life of an installer can be, there will come a time when you get bored or tired or maybe even ambitious enough to start looking around you to find new opportunities, and new adventures. Certainly there are plenty out there. You may have your life planned out and know exactly what you are going to do and when. I think that most of us have yet to figure that out though.

The first step is to identify what it is that you like, or find interesting about your job. Is it that you like the people part of it, or the paperwork (some guys like that part of it, really!)? Or maybe it's the physical challenge or that you like working with the high tech equipment. What about working with the other techs themselves or teaching the helpers? The traveling or working in a new place is fine, too. When what part you enjoy most becomes clear, then you can begin to plan the next step.

If during your turn as an installer, or service tech, you show true interest in what you do, the next step may come to you before you even start looking. If you are especially good at getting people to do things your way, your co-workers respect you, and you seem to know what you're doing and are concerned with doing it properly, the next logical step may be supervisor, or project manager. If this is your goal, let the higher ups in your company know, and don't be shy about it. If you're going to hold this position, you'll need to learn how to project authority and confidence, while not coming across as the class bully. It's a fine line that comes naturally to some people. If this sounds like you, start by asking to run a larger job, with more people than you usually work with. It's not always easy as it looks, but can be just what you're looking for. Take that one step farther, and that may bring you to department head. This position becomes less technical and more managerial. By then you don't want to be in an attic anymore, anyway. You'll spend more time planning jobs, and how to improve the department's profitability, rather than what thermostat to use.

If you really like the technical end of it, and can't wait to see the new cross-linked, titanium, hydro-injected, posi-traction, Acme 2000 38 SEER condenser, then you may be just the guy to hook up with the manufacturer, and get on the equipment representative train. These people are the ones giving out the free pens at the company barbeque. They travel often, and generally don't get dirty. Their job is to convince you, or your boss, that the do-hickey that they represent is better than the don't-hickey you're currently using. They don't usually sell the product directly, only the concept. They need to build a relationship with your company, and show how ready they are to help. They do trade shows, maybe some home shows, and mostly talk about their product. The money is usually good, and not a whole lot of stress there, unlike the management positions.

The other factory job you may come across is that of a factory tech. Here you'll be the local expert or brand X equipment. You'll have access to the factory guys that design the stuff. Again, you'll travel quite a bit in your territory and have a company vehicle. One of the nice things about being a factory tech is that everyone wants to talk to you. However, one of the bad parts may be that everyone wants to talk to you. You decide. If you work for a good company, you will be the hero riding in to fix the problem nobody else could figure out. If not, well, learn to shrug really well, and carry a lot freebies.

The next possibility is sales. Wait, wait come back here and sit down. You don't need to be the greasy used car salesman that we all think of.

Sales in this field can be a lot fun, and very rewarding. In fact I'm going to encourage you to give this a try, no matter what direction you really want to go. Most companies will continue to send you for more and more training, you'll meet a lot of people, and you're never stuck in one place for very long.

Not only are good salesmen always in demand, but did you know that salespeople are the highest paid people in the world? Very often you'll be able to make $100.00 to $1000.00 per sale, or much more. Don't get me wrong, if it were truly that easy, nobody would ever do anything else. But there are an awful lot of people who took a job in sales because that was all they could find at the time, only to find that it was the greatest move they ever made. It's not uncommon in the world of sales to find a high school dropout making a very nice six-figure income. I know many. Once you get good at sales, you can go anywhere. For now, start with a field that you know well, and don't be nervous about the commission part of it. If you choose to take a salary, you not only have no real incentive to become something great, but you are giving all the real money to your boss. Believe me, commission is best for almost everyone in the long run. Oh, and remember, everything is negotiable, including your compensation.

Now then, if you think you can do it all and more, and do it well, then you may want to start, or buy your own company. This is always, not sometimes, but always, much more involved than you can even believe. I am definitely not saying it is bad; only that you need to be ready to have this project completely engulf your life for years to come. I have never heard someone say, "Yeah, it's actually easier than I thought." No sane owner has uttered those words in all the history of the world. Ask one. They'll probably offer to sell you theirs. The only problem is that they won't ever sell it for what it is actually worth, because they have five times more than that invested in it. However, if you are careful, and do it right, you can reap great success, and pride. There are so many rewards with being self-employed that people will continue to give it their all, to try to do it better than the other guy. That's what makes this country so great. Today you are a helper, tomorrow you own the company.

Well, it's been fun. I wish you happiness and good fortune with whatever direction you ultimately choose. I believe that if you took the time to read the entire book to the end, you already have enough of what it takes to do well in any of these post tech occupations. Good luck, stay open-minded, and WIN!

CPSIA information can be obtained at www.ICGtesting.com
Printed in the USA
BVOW031821181211

278669BV00006B/128/P